Elen Silva de Andrade
Lurdes C. V. C. de Garay

Innovative teaching methods in Geography

Elen Silva de Andrade
Lurdes C. V. C. de Garay

Innovative teaching methods in Geography

Project methodology in geography teaching

ScienciaScripts

Imprint
Any brand names and product names mentioned in this book are subject to trademark, brand or patent protection and are trademarks or registered trademarks of their respective holders. The use of brand names, product names, common names, trade names, product descriptions etc. even without a particular marking in this work is in no way to be construed to mean that such names may be regarded as unrestricted in respect of trademark and brand protection legislation and could thus be used by anyone.

Cover image: www.ingimage.com

This book is a translation from the original published under ISBN 978-613-9-73153-4.

Publisher:
Sciencia Scripts
is a trademark of
Dodo Books Indian Ocean Ltd. and OmniScriptum S.R.L publishing group

120 High Road, East Finchley, London, N2 9ED, United Kingdom
Str. Armeneasca 28/1, office 1, Chisinau MD-2012, Republic of Moldova, Europe
Printed at: see last page
ISBN: 978-620-7-88583-1

ACKNOWLEDGEMENTS

Firstly to God Almighty, who in his greatness has guided us, giving us the strength to achieve our goals and overcome the difficulties that have arisen along the way. And for granting us the grace and wisdom to produce a work of this magnitude.

To my supervisor, Professor Dr Lurdes Célia Virginia Calderini de Garay, for her understanding and support in carrying out the work presented here.

To the teachers at Mineko Hayashida school, especially the geography teachers, who were part of the construction of this work, contributing their pedagogical concepts.

To my friend Elizabete Rodrigues, for her tireless support in monitoring this work.

To the students who participated in the development of this work and contributed by answering the questionnaires and making them a reality.

To my family, who have always encouraged me and have always been by my side, especially my parents Belarmino Silva de Andrade, for their dedication to the family and in memory of my mother Maria Raimunda de Andrade, who could not see the realisation of this dream that is now coming true, and to my three beloved children for their patience and tolerance in the absent hours of their lives.

To the friends who directly and indirectly encouraged me on this journey and believed in my dreams, as they were extremely important.

SUMMARY

In the 21st century, geography is still taught in a static way, based on traditional classes. Given this problem, it is not appropriate to use these practices at the present time, since geography itself is an interdisciplinary science. The aim of the research: Teaching Geography in Contemporary Times: New Demands and Innovative Didactic Possibilities, Laranjal do Jari - Amapá, is to analyse the practices of geography teachers at Mineko Hayashida State School and identify whether the methodologies used are traditional or innovative. In order to carry out this work, a questionnaire was used as a data collection tool to answer this question. After administering the questionnaires to five of the teachers surveyed, it was found that they use dialogue-based lectures and that sporadically, in addition to textbooks, they use the media, since the school does not have an adequate structure for innovative teaching. To the detriment of the aforementioned resources used by the teachers, it was observed that this methodology was not enough to develop the teaching and learning process in a critical approach. In order to minimise this problem, the pedagogy of projects, a technique advocated by Frenet, was used with 29 students taking part in the research. The result of this technique enabled lessons that went beyond the classroom, such as multimedia and practical lessons in the field, facilitating learning, as well as forming students into complete and critical citizens capable of acting as authors of their own history.

Keywords: Practices; geography teaching; innovative methodologies

SUMMARY

CHAPTER 1

INTRODUCTORY FRAMEWORK

1.1.   INTRODUCTION

This research has become relevant because it proposes new methodologies for the teaching and learning process, both for students and teachers. In the 21st century, it is not appropriate to work on Geography only in the classroom, but to encourage students to instigate their curiosity in the search for the unknown and to help them organise or systematise their ideas and thoughts.

The classroom environment in the subject of Geography should be very interactive, where teacher and student build knowledge together, each subject collaborating in this process, thus making teaching and learning positive, and then it should be remembered that this knowledge acquired at school should be put into daily practice in society or vice versa.

There is no doubt that the subject of Geography is a science that contributes to a better understanding of the world in which we live, and we know that this world is in permanent transformation; transformations that can be of a natural and/or social nature.

The role of the teacher/guidance counsellor is to enable pupils to act as citizens, making decisions and taking action in the society in which they live, so the aim of education is to take into account all the potential of each individual, such as: memory, reasoning, aesthetic sense, physical abilities and the ability to communicate.

This is how education is conceived that forms the citizen as a whole, which can be understood and defined in the development of education programmes and new pedagogical policies. In this sense, forming complete citizens in accordance with the four pillars of education, capable of sensitising themselves to the different situations of everyday life, is something that the world lacks in human beings.

The research was structured as follows:

The theoretical framework is the second chapter of this work and is subdivided into the following seven items: the history of school geography in Brazil; understanding geographical science;

a brief history of pedagogical trends: human development and learning theories; teaching geography in contemporary times: new demands and innovative didactic possibilities; innovative didactics; the school environment as the acquisition of knowledge; interdisciplinarity as a proposal for change and innovation in the teaching and learning of geography.

The third chapter of this research deals with the methodology applied and its instruments and techniques for obtaining the desired data. The methodology of the research under study is qualitative and qualitative, where the actions developed from the application of questionnaires to students and geography teachers, in order to analyse, compare and interpret the data of the current study.

The fourth chapter deals with the analytical framework, i.e. the object of the current study, based on the analyses of the data collected.

The fifth chapter includes conclusions, recommendations, references, annexes and appendices.

This study took place in 2015 at the Mineko Hayashida state school, which operates only as a secondary school, located at Av. Tacredo Neves, 2960, Agreste, Laranjal do Jari - Amapá - Brazil.

In order to carry out the practical activities in the classroom, the students from six first-year high school classes were divided into two groups: the first group was made up of 29 students who had innovative lessons during one school year, and the second group was made up of 211 who received traditional lessons, both with the same workload of 80 hours per year, divided into 4 bimesters.

Next, data was collected from the questionnaires applied and observation of the activities applied in group 1, in order to discuss them. The results of the research between the two groups of students were compared and evaluated, as were the practices of the geography teachers.

1.2.   Theme

Teaching geography in contemporary times

1.3.   Title

Innovative teaching methods used by teachers to teach geography at secondary level at Mineko Hayashida State School, AP - Brazil.

1.4.   Problem description

In the 21st century, should we still insist on teaching geography in a traditional way without innovative practices?

In his work, Rego (2007) says that teaching geography means "merely exposing a programme of content, assumed to be invariably already established, accompanied by an assessment of its assimilation by the students". Based on this statement by the author, the problematic situation in current geography teaching occurs because it is still uninteresting and also because it still focuses on memorising content. This practice, with its retrograde method of passing on content to students, needs to change, because the content that forms the curriculum, the basis of the secondary school curriculum, needs to be related to everyday life. This change also includes presenting content in a more contextualised and dynamic way, forming a connection between theory (content taught in the classroom) and practice (student experience). According to Cavalcanti (2010), although geography teaching has made great theoretical progress, this progress has reached school practice very slowly or belatedly and this slowness means that current teaching is still based on traditional methodological practice.

What can be seen in geography classes is that it is difficult for students to produce their own knowledge by producing texts or even analysing poems, graphs and other languages in the field of geography. This difficulty can be explained by the way teachers teach and their pedagogical practice, by the lack of geographical literacy in teacher training and, as a result, this problem persists in secondary education. According to Castrogiovanni (2000), "this is because the teachers who work in the initial grades have not been literate in geography", so students will arrive at secondary school conditioned to geography teaching with a deficiency in its basis.

Thus, based on these difficulties, a traditional teaching that was once conceived is referred to or considered, but Geography as a dynamic science, it is not convenient to continue with this obsolete Geography teaching. The objective of teaching geography is still a challenge for teachers. This challenge is to increase the relationship between teacher and student and to make geography lessons an effective reflection on the themes studied, producing changes in the student's way of thinking and acting, through innovation in their methodological practices so as to improve the quality of geography teaching.

1.5.   Questions

1.5.1. General

What innovative didactics do teachers use to teach Geography at secondary level at Mineko Hayashida State School in Laranjal do Jari - Amapá - Brazil 2015?

1.5.2. Specific

1.5.2.1.   What innovative didactics do teachers use to teach Geography at secondary level at Mineko Hayashida State School in Laranjal do Jari - Amapá - Brazil 2015?

1.5.2.2.   What strategies do geography teachers use at Mineko Hayashida State School in Laranjal do Jari - Amapá - Brazil 2015?

1.5.2.3.   What is the need to apply new methodologies in teaching Geography at Mineko Hayashida State School in Laranjal do Jari - Amapá - Brazil 2015?

1.6   Objectives

1.6.1   General objective

To analyse the innovative didactics used by teachers to teach Geography at secondary level at Mineko Hayashida State School in Laranjal do Jari - Amapá - Brazil 2015.

1.6.2   Specific objectives

1.6.2.1.   Describe the praxis of high school geography teachers at Mineko Hayashida State School in Laranjal do Jari - Amapá - Brazil 2015.

1.6.2.2.   Demonstrating the methodological strategies used by the geography teacher at Mineko Hayashida State School in Laranjal do Jari - Amapá - Brazil 2015.

1.6.2.3.   Point out the need to apply new methodologies in the teaching of Geography at Mineko Hayashida State School, in Laranjal do Jari - Amapá - Brazil 2015.

1.7   **Justification**

This research has become relevant because it proposes new methodologies for the teaching and learning process, both for students and teachers.

If we only work on Geography in the classroom, we must instigate our students' curiosity in search of the unknown and help them to organise their ideas and thoughts.

The classroom environment in the subject of Geography should be very interactive, where teacher and student build knowledge together, each subject collaborating in this process, thus making teaching and learning positive, and later, we must remember that this knowledge acquired at school must be put into daily practice in society.

Without a shadow of a doubt, Geography is a science that contributes to a better understanding of the world in which we live, and we know that this world is in a state of permanent transformation; a transformation that can be of a natural and/or social nature.

The role of the teacher/guidance counsellor is to enable students to act as citizens, making decisions and taking action in the society in which they live, so the aim of education is to take into account all the potential of each individual, such as: memory, reasoning, aesthetic sense, physical abilities and the ability to communicate.

This is how education is conceived that forms the citizen as a whole, which can be understood and defined in the development of education programmes and new pedagogical policies. In this sense, forming complete citizens in accordance with the four pillars of education, capable of sensitising themselves to the different situations of everyday life, is something that the world lacks in human beings.

Nowadays, through the technical-scientific era, it is not appropriate to work in a watertight manner, without taking into account the transformations in today's society.

In this respect, Rego (2007, p. 41 and 42) says that deconstruction at school involves a return to interdisciplinarity, which is still a challenge for teachers, as it requires greater commitment and dedication, discovering no longer means uncovering something that was covered up by reality, but inventing new relationships between scientific concepts. This position is part of modern (post-modern?) paradigms, but it is known that "the majority paradigms still privilege the space of specialisation and the frontier between knowledge, which overrides creative work".

Geography as a subject in the school curriculum is unfortunately still seen as uninteresting by many, perhaps because it's a geography that focuses on the study of rivers, vegetation, relief, countries, etc.

Or just a subject for memorisation, culturally explained by traditional teaching. However, in the first half of the 21st century, Geography, more than ever, places human beings at the centre of its concerns, which is why it can also be considered a reflection on anthropic actions in all their dimensions. It is concerned with the concerns of today's world, seeking to understand the complexity of how order and disorder occur on the planet. In reality, it is an instrument of power for those who possess its knowledge.

In these respects, it is extremely important that we break old paradigms and find new ways of attracting our students' attention to the study of this science through a reflective, logical and critical geography, and for this we need a re-signified, contextualised curriculum where our students can feel that they are actors in their own historical construction process, rather than the passive agents of a traditional education.

As this research is focused on education in the search for innovative methodologies, it will be carried out through participant observation and may use qualitative research.

1.8.   Research limitations

This research aims to observe students in the first year of secondary school and the practices of geography teachers at Mineko Hayashida State School. The problem of lack of interest in geography classes is recurrent and is due to poor teacher training. Knowing that geography is a dynamic science, it enables the interdisciplinary study of other sciences, making the teaching-learning process more contextualised. Based on this idea, we realise the need for a methodology that meets the dynamism of geographical science, given that we live in today's scientific and informational technical society, and that this technological society tends to be in constant transformation. Thus, the main limitation is related to the theme: Teaching Geography in Contemporary Times, where it should lead students to reflect on the issues of their time in the local, regional and global spheres. Thus, this knowledge acquires new meaning in the student's experience.

CHAPTER 2

THEORETICAL FRAMEWORK

2. 1 THE HISTORY OF SCHOOL GEOGRAPHY IN BRAZIL

During the sixteenth, seventeenth and eighteenth centuries, the Brazilian educational process was governed by the Jesuits, who differentiated between the indigenous people and the children of the settlers. For the indigenous people, this education turned to religion in order to evangelise them, while for the administrators and settlers, it was already an education of a humanist nature, in other words, "love of country" and in a "poetic and romantic form of the landscape". In this way, geography was taught in the form of literary texts.

And in the 19th century, during the Empire and the Republic, education in Brazil prioritised the privileged, wealthy class: intellectuals, military professionals, civil servants, small traders and artisans.

However, in 1931, geography became a requirement for admission to higher education law courses, thus training administrators for public office in Brazil.

However, the institutionalisation of geography in Brazil took place at Colégio Pedro II. According to Cristódio (2010, p.2) apud (Bordieu, 2004), institutionalised knowledge means that it is relevant to influential social groups. In this sense, geographical knowledge becomes a tool for ascending to power.

According to Calvanti (2010. p.4-5) "the geography renewal movement" (academic and school) in Brazil began in the 1980s. Its beginnings were marked by a dominant competition between two main centres: one concentrated on the so-called "traditional" geography, which had maintained its structure since the first decades of the 20th century; and the other, which addressed a new geography that sought to superimpose traditional geography, thus promoting itself as critical geography.

Faced with this crisis, the aforementioned nuclei realised that the purpose of critical geography was to denounce the utilitarian character and ideology of geographical science, as well as the:

the false neutrality and innocence of "official" geographical thinking and geography in the classroom. The aim was to advance the understanding of space, its historicity and its dialectical relationship with society. From the first period, there was a diversity of understanding of what critical geography was... This diversity was consolidated in the 1990s. (Calvanti, 2010, p.5)

According to this perspective, it is possible to observe a fragmentation in what is taught in geography in the classroom.

In 1980, the geography "movement" was also centred on giving greater emphasis and social significance to the school subject under study.

However, from the 1990s onwards, according to Cavalcanti (2010), "in the socio-political, scientific and educational context and crisis" in teaching geography, more articulate proposals were redefined from the didactic-pedagogical point of view, based on teaching methods with geography. In the face of these changes, we recognise those of an everyday nature, in which this society..,

globalised, urban, informational and technological, require an understanding of space that includes subjectivity, everyday life and multi-schooling,

the different languages of today's world (Calvacanti, 2010, p.5).

However, in the 1990s and 2000s, it was consolidated in undergraduate and postgraduate courses, and in a network with primary school teachers.

2.2 GEOGRAPHICAL SCIENCE

2.2.1 A brief history of geographical currents

Traditional geography, according to Andrade (1987, p. 67). Geography focused on astronomy and cartography studies, where its main functions were to organise routes and study natural resources in order to exploit them.

In the Modern Age, there was a legitimate advance towards economic, social and cultural development, with a new perception of the relationship between society and nature and a new look at the object of study of geography, which, based on this perception, sought to explain certain natural phenomena and redefine the concept of geographical space. It wasn't until the 19th century that geography became a science, with its main contributors being Humboldt (a naturalist and traveller) and Ritter (a philosopher and historian) and, at the end of the 19th century, their followers Ratzel, who advocated a colonial empire for Germany, and Elisée Recleus, who developed a libertarian theory and condemned the expansion of colonisation.

In Ratzel's vision, through traditional geography, space is conceived as fundamental in the life of man, so it will serve the nation state, taking care of governments through situations

In the mid-twentieth century, schools were committed to governments and depended on the state, i.e. they were subordinate to its wishes:

> In Germany, to justify and try to legitimise the struggle for living space, in France and Great Britain to better understand their colonial empires, in the United States and Russia to justify and consolidate expansion into continuous areas inhabited by poor peoples who would remain under their domination and guidance. (Andrade, 1987, P.67)

This legitimacy of space as fundamental to man led geographers to analyse part of the space with a view to knowing the whole. Geography was thus consolidated as a science of description, in other words, descriptive and ideographic. As a science of description, man was not part of the process.

However, the descriptive phase of geography has left its legacy as a traditional phase through concepts such as habitat, environment, territory and region, which are being discussed today.

2.2.   2Theoretical-quantitative geography (Correa, 2003, p.19)

Theoretical-quantitative geography is also known as New Geography, which used mathematical and statistical models during the post-war period, culminating in technological advances. This geography catered for authoritarian governments and large economic corporations, which, by using purely quantitative data, did not take into account spatial, social and cultural peculiarities.

their attitude was only for the enrichment of the capitalists.

In this geographical current, a new geographical perspective emerges, where the concept of space is re-signified, which according to Corrêa (2003, p. 22-23):

> This is a limited view of space, because on the one hand, distance, life as an independent variable, is over-privileged. On the other hand, contradictions, social agents, time and transformations are non-existent or relegated to a secondary level. An eternal present is favoured and underlying this is the paradigmatic notion of (spatial) equilibrium, dear to bourgeois thinking.

Analysing geographical space only from a quantitative-theoretical point of view simply means analysing numbers and statistical data, i.e. quantitative analysis of the complexity that exists in space. However, in the light of the above, Corrêa says that this current has brought contributions to geography in terms of the possibility of extracting knowledge about locations and flows, hierarchies and functions. "Such models provide us with clues and indications that are effectively relevant to a critical understanding of society in its spatial and temporal dimension, and should not be considered as normative models as they were intended to be" (Corrêa, 2003, p.23).

In addition, in school, geography teaching in this current appropriates the use of statistical data, disregarding field lessons and the observation of the reality experienced by the student.

2.2.1   Critical or radical geography (Andrade, 1987, p.122)

This geography emerged in the 1970s in the midst of an economically, socially and politically turbulent period of unbridled pursuit of wealth, the main source of which was nature and its natural resources. At the same time, social inequalities intensified, and a political movement arose among the popular masses, demanding and proposing a reform of society.

As all events are interconnected, what emerged from the countryside and the popular masses was reflected in the scientific field, and this situation led geographers to analyse the relationship between man and nature in order to understand the phenomena of the reality of that moment.

However, this critical aspect of geography was treated with scientific rigour, and received a major contribution from the geographer Milton Santos, who discussed the concept of space in depth in his works, establishing categories that best analyse this space: form, function, structure and process.

Santos (2002, p. 19) considers that "form" is the visible aspect of the object (materiality); "function" is an activity or task of that object; while "structure" is linked to the social and economic nature of a society, and "process" is considered to be an action of continuity, involving time.

2.2.2   Geography in secondary school

Geography teaching must start from the premise that knowledge does not occur separately as if it were divided into drawers, so in secondary education, it must seek or adapt to a comprehensive vision, that is, various ways of thinking or teaching certain content by relating it to other sciences, in this way

Preconised by the PCNs where there are:

> A desirable and already existing context has expanded the participation and debate of teachers and students in discussions and the teacher is no longer a mere transmitter of knowledge, but thinks about the world in a dialectical way. This process has opened up the possibility of effective methodological integration between the different areas of knowledge and Geography, from an interdisciplinary perspective. (PCNs, 2008, p. 44)

Interdisciplinary teaching in the subject of geography allows for a broad formation of citizens in the way they think or reflect on their actions in society, actions that are based on the reality they experience.

School geography (is there more than one geography?), according to Rego (2007, p. 43), more than ever, must be taught in such a way as to equip students to deal with spatiality and its multiple approaches: where they can operate space! This approach seeks to understand social life as reflected in the different subjects responsible for (trans)formations. This makes it easier for the subject to recognise social contradictions and conflicts and to evaluate the appropriation and organisation established by social groups.

The act of reflecting on space is not an easy task, since geography is a dynamic science and such dynamism is due to the possible tools for analysing geographical space, and this analysis is only possible through the categories of analysis studied as a form of geographical literacy.

Geography teachers need to be constantly reflecting on what geography is and what it should be, since it is not a static science, but also a school practice, a life practice, a practice that students experience.

Given the above, geography is perhaps the discipline that works most with interdisciplinary practices, as it opens up a range of possibilities in the field of education. In the globalised world, there is no avoiding the use of geography's concepts or categories of analysis - place, region, landscape, territory, territoriality - in order to understand the different conceptions of the world in which we live.

According to the PCNs (2006, p. 46), one of the great challenges for geography teachers is to select content and create strategies on how to work with it in the classroom. These challenges are caused by the rapid transformations taking place in today's world, especially in the technological sphere.

Faced with this scenario, the geography teacher plays an extremely important role in everyday school life, in creating strategies that can improve teaching and learning and lead them to acquire the ability to analyse their reality from the geographical context, and for an analysis of geographical facts or phenomena there is a need for written production.

With the evolution of technology as a means of opening up a new direction for education, the aim is to improve learning in practice for broad knowledge between student and technology. Olson (1976, p. 18) writes:

> The invention of cultural devices, instruments and technologies that include invented symbolic forms such as oral language, writing systems and numerical systems,
>
> Iconic resources and musical productions allow and demand new forms of experiences that require new types of skills or competences.

Language, culture and technology always go hand in hand, one complementing the other in a way that makes it possible to analyse how to understand both in order to understand the uses that these skills will be put to in the classroom.

> [...] "Writing as technology" is what has led me to understand how the production of knowledge about language comes from the construction of the subject's knowledge about themselves and about the world, because at the same time that the subject produces and practices a technique to tell themselves and relate to others, they produce a "stirring" in the structure of language. We know that language and culture are not separate, so the moment the subject is affected by the meanings of a given technological culture (ideology), there are necessarily repercussions on language. [...] (Orlandi, 2001, p. 40)

With so many ways to find out in this new age of computerisation, it's even difficult to practise the standard teaching required in schools.

Today, in our cities, most education takes place outside of school. The amount of information communicated through print, magazines, films, television and radio far exceeds the amount of information communicated through instruction and texts at school. This challenge has destroyed the monopoly of the book as an aid to teaching and broken down the walls themselves so suddenly that we are confused, bewildered (Carpenter, McLuhan 1960).

2.3 CURRICULUM PARAMETERS   FOR GEOGRAPHY TEACHING

The purpose of this topic is to discuss the importance of the PCNs for secondary education in the field of Geography, serving as a tool for

to enrich teaching and help geography teachers reflect on their daily methodology.

2.3.1 The   importance of the PCNs in teaching geography

According to the PCNs (2006):

The reformulation of the curriculum in Brazil took place through the National Curriculum Parameters (PCNs), where it was defined that secondary education was divided into areas of knowledge, namely: language, codes and their languages, human sciences and their technologies, natural sciences, maths and their technologies.

This reorganisation of the curriculum into areas of knowledge aims to make it easier to develop content, with greater possibilities for interdisciplinary study and contextualised teaching.

Still in line with the PCNs, geography in particular is divided into three competences: representation and communication, research and understanding, and socio-cultural contextualisation.

For representation and communication, this competence is

> "constitute the basis for defining the principles of documentary analyses and procedures that are fundamental to the exchange of information and records between the various disciplines of the area and other areas." PCNs, 2006, p. 59)

In this geographical competence, both teacher and student are encouraged to research, thus building scientifically-based principles or notions, enabling the development of techniques to record and document such research and thus become the communication and exchange of information in related areas.

The pupil in secondary school will thus have within the representation and communication in geography will be given like this:

> "Read, analyse and interpret specific codes (maps, graphs, tables, etc.) considering them as elements of representation of spatial or specialised facts and phenomena". PCNs, 2006, p. 59)
>
> "Recognise and apply the use of cartographic and geographic scales as ways of organising and knowing the location, distribution and frequency of natural and human phenomena." PCNs, 2006, p. 59)

With the appropriation of the skills mentioned above, it is possible to work with the geographical space, identifying phenomena caused by anthropic action, with the instruments present, it is possible to analyse through mathematics elements that make up the location, such analysis is done through "maps".

Tables and graphs allow the teacher and student researchers to get a sense of the dimension of space and reality and its representativeness. In order to produce documents using the elements mentioned above, it is necessary to know certain technologies in order to achieve the expected result.

The second competence, research and understanding, involves working on scientific geographical research practices. From this perspective, the teacher is able to raise or list topics from the student's daily life and systematise the work with a scientific character. Competences developed from this perspective are:

> "Recognise spatial phenomena through selection, comparison and interpretation, identifying the singularities or generalities of each place, landscape or territory." (PCNs, 2006, p. 59)
>
> "Selecting and drawing up research schemes that develop observation of the processes of formation and transformation of territories, with a view to labour relations, the incorporation of techniques and technologies and the establishment of social networks." (PCNs, 2006, p. 59)
>
> "To analyse and compare, on an interdisciplinary basis, the relationship between the preservation and degradation of life on the planet, with a view to understanding its dynamics and the globalisation of cultural, economic, technological and political phenomena that have an impact on nature, at local, regional, national and global scales." PCNs, 2006, p. 59)

These competences make it possible to identify phenomena and their location, being able to study them in the most varied forms and complexities of the science under study, considering instruments capable of comparing and interpreting phenomena according to their peculiarities and geography's own technologies as tools to make study relevant.

The appropriation of the above aspects subsidies studies of the phenomena and/or actions caused by man's actions, where such studies can minimise the impacts caused on all scales, given that "thinking local is acting global".

Another group of competences in geography and the last one, according to the PCNs (2006, p.60)

> "Recognise in the appearance of the visible and concrete forms of today's geographic space its absence, that is, the historical processes constituted at different times and the contemporary processes, the set of practices of different agents, which result in profound changes in the organisation and content of space". (PCNs, 2006, p.60)

In view of the above, the possibility for an interdisciplinary practice is elucidated, specifically in the process of teaching and learning geography.

The competences and objectives of Geography for secondary education can thus be summarised as basic, according to the PCNs:

> "Reading and interpreting cartographic documents (maps, graphs, tables), as well as drawing them up." (PCNs, 2006, p.61)

> "Identifying and interpreting the constituent structures of geographical space in its various units".

> "Recognising and identifying the constituent elements of geographical space, including evaluating their incorporation into the process of production/appropriation of geographical space." (PCNs, 2006, p.61
> "Evaluation of its impacts, both from a historical perspective and in relation to the present moment." (PCNs, 2006, p.61)

2.3.2 Curriculum reformulation in geography teaching

2.3.2.1 Defining the curriculum

Addressing the concept of school curriculum for various experts on the subject will clarify or remind us of its meanings and functions in the school environment.

Dealing with this subject requires an understanding of how human beings have produced and are producing in the course of their history, productions that are transformed over time.

For (Cunha, 1986, p. 235), curriculum has its Latin origins in curriculum, which means "the act of running, a shortcut, a cut". The author makes an analogy, comparing the curriculum to a path taken by boats on a river. The boat follows its route, facing bends, rapids and snags. As it goes along, each boat makes its way, facing adverse physical, emotional and cognitive conditions, among others.

From this analogy, it can be seen that the curriculum goes through various emotions or issues that accompany the whole process. In this way, it is necessary for the path or curriculum to take into account the authors involved, taking into account the knowledge and experiences of those who take part in this process.

So the curriculum consists of transformation, which is not just about changing, or going down other paths, but also seeking new alternatives, solutions and achievements, where this search will result in the quality of teaching and learning in the educational process.

According to the Aurélio Dictionary of the Portuguese Language (Ferreira, 1986, p. 512), the curriculum is defined as "the part of a literary course; the subjects contained in a course". In this sense, the curriculum is presented as just a list of contents taught in the disciplines of related courses.

Each author contributes in a different way to the concept and function of the curriculum in schools, Sacristã (2000, p. 14) states that:

> The curriculum as a set of knowledge or subjects to be mastered by the student within a cycle - educational level or modality of teaching - is the most classic and developed meaning; the curriculum as a programme of planned activities, duly sequenced, ordered, methodologically as shown in a manual or teacher's guide; the curriculum has also sometimes been understood as intended learning outcomes; [...].

From the attributions to the curriculum, it is possible to see that it is related to the competent bodies in drawing up a list of content to be complied with by the school [...]. However, Zotti (2008) states that the curriculum goes beyond the list of content, taking into account social, ideological, cultural, political and historical issues, issues that permeate the educational process that are not taken into account when drawing up the curriculum.

Still on the subject of conceptualising the curriculum, Sacristã (2000, p. 14) states that:

From this perspective, it is necessary to understand that in order for students to retain the knowledge developed by human beings historically, they need to have a critical understanding of the reality they experience, and not as a repetition of content that has been covered without contextualising it.

According to Saviani (2003, p. 23), "the set of activities developed by the school. From this point of view, the curriculum is everything the school does, so it wouldn't make sense to talk about extracurricular activities".

2.3.2 Curriculum as Social Approaches

For Sacristã (2000, p. 14), "the curriculum as a programme that provides content and values for students to improve society in relation to its social reconstruction". From this point of view, the school is the institution with the technical capacity to provide selected knowledge and the students, in possession of this knowledge applied according to their experiences, would be able to take part in the process of social transformation, where they would participate as actors in their own history.

When the school doesn't use the process of democratising the curriculum, or doesn't apply it taking into account social, ideological, cultural, political and historical issues, Soares says that:

By denying the working classes the use of their own language (which it censors and rejects), while failing to lead them to master the language of prestige, the school is fulfilling its role of maintaining discrimination and marginalisation and therefore reproducing inequalities (Soares, 1989, p.72).

Thus, the author is not saying that the school should not follow the formal standards of how to teach grammar, for example, but that the school should look for strategies to teach these grammar rules, giving due meaning and applicability of these rules to the student's daily life.

Saviani (2003, p. 23) talks about curriculum, which we believe in:

The school exists, therefore, to promote the acquisition of tools that enable access to elaborate knowledge (science), as well as access to the very rudiments of this knowledge. Basic school activities must be organised on

> this basis. If we call this the curriculum, then we can say that it is on the basis
> of systematised knowledge that the primary schools curriculum is structured.
> Systematised knowledge, erudite culture, is a literate culture. Hence the first
> requirement for this type of knowledge is to be able to read and write. You
> also need to learn the language of numbers, the language of nature and the
> language of society. This is the fundamental content of primary schools:
> reading, writing, counting, the rudiments of the natural sciences and the
> social sciences (history, geography, humanities).

In this sense, the curriculum is understood as the production of the actors involved in the teaching and learning process, as it is characterised as a production practice and not as a bureaucratic document. It requires an incessant search for systematised knowledge and an effective understanding of and belonging to the socio-economic and cultural reality that influences the educational process.

According to Neder (2004, p.21), using the work of Costa (1999), the curriculum must be understood as a social production that is situated in a correlation of forces that determines criteria of validity and legitimacy through which representations, meanings and realities are produced and instituted. [...]

According to the author, it is a place where narratives circulate, but above all it is a privileged place for the processes of subjectivation and directed socialisation.

The democratisation of the curriculum and the school can also take place in the drafting of the school's Pedagogical Political Project, in which everyone can discuss their positions, points of view and proposals. But for this to happen, the management team and teachers need to respect the positions, points of view and proposals of parents and students. This means accepting that they have a lot to tell us and that, if the teaching-learning process seeks knowledge and this must be realised for the students, those directly concerned must be able to have their say and make proposals.

A second important point is the change in the teacher-student relationship. For a school to be truly democratic, this relationship cannot be based on one of command and obedience. The teacher cannot see the pupil as someone devoid of knowledge, because when the pupil arrives at school, he or she brings with him or her the knowledge he or she has experienced in the family and in society in general. We don't want to echo the current neoliberal proposals which, taking the above as their starting point, turn the teacher into a mere observer of the student's learning. We don't agree with this view.

The teacher plays an extremely important role in the teaching-learning process. It is the teacher who has the basic knowledge of the subject being taught and also of teaching techniques. What we want to emphasise here is that in order for there to be real democratisation of the school, the teacher must see the pupil as the builder of his or her history and knowledge, and it is essential that the teacher listens to his or her pupils and participates with them in the teaching-learning process.

In order for this to happen, we should look for a process that starts from the students' knowledge and progresses from there to the maximum knowledge possible. Starting from what the student knows doesn't mean starting from the individual, but from the group or class to which they belong, and it also doesn't mean sticking to what the group already knows. As stated above, the search for ever higher knowledge should always be the goal of the teaching-learning process (Saviani: 2000).

2.4 A BRIEF HISTORY OF PEDAGOGICAL TRENDS: HUMAN DEVELOPMENT AND LEARNING THEORIES

According to Lakomy (2003, p.12), "there are various metaphors for what learning is". Let's look at a few: in the metaphor of the container, the student's mind receives a range of information about reading in books, hence the idea of learning. In another metaphor, learning can occur through the accumulation of knowledge, where such knowledge occurs at levels 1, 2, 3, 4 and 5. Starting from these levels of knowledge, it is enough to fill the student's mind with new knowledge for them to learn.

For the author, "we can't get rid of metaphors altogether", because in our daily lives we are confused by external manifestations and products.

These are metaphors for explaining learning, but Lakomy (2003, p.16) conceptualises it as learning, apprehendere is, to apprehend with me, is to become for oneself, or to become "one's own".

Thus, learning is the acquisition of behaviour that the subject did not previously display.

2.4.1. Theories of Piaget and Vygotsky

For the Swiss Jean Piaget (1896-1980), the father of the psychology of intelligence, who was concerned with how we learn something? He turned to experiments with children based on open dialogues on specific topics rather than tests. Through these tests, he sought to understand the evolution of basic concepts such as: space, time, physical chance, number, moral judgement, etc.

Through his studies applied to his own children, Piaget defined a new concept: constructivism, where learning is actively constructed by the subject themselves, it is a consequence of their interactions with the world and their reflections on these experiences, of everything they can abstract from them. Piaget says that children are like "little scientists", since they actively experiment with the world.

After a talk to pre-school teachers, Piaget exaggerated his words by saying that "language is useless". His obsession was to explain intelligence as a biological process. In his autobiography, published in 1971, he tells how, as a teenager, he had a "revelation" reading a philosopher named Bergson: "In a moment of enthusiasm bordering on ecstasy, I was seized by the certainty that God was life, in the form of a vital impulse, of which my biological interests provided me with a small sector of study. [...] This led me to take the decision to devote my life to the biological explanation of knowledge."

In Piaget's theory of learning, the interaction between the subject and the object is taken into account, disregarding the social exchanges in these senses. Palagana (2001, p. 162) states that:

> [...] focuses on the subject, seeing the object only as a potentially disruptive element in the cognitive structure. In this way, in Piagetian constructivism there is no reciprocal exchange, no equal influence between the two poles of the unit of knowledge, which characterises the very nature of the interactionist approach.

In view of this, we realise that we cannot disregard social exchanges, since we live in society, and as social individuals our interactions become generators of learning in its various forms.

In contrast to Piaget's ideas, Vygotsky, according to Andrade (2011, p. 260), says that social interactions are the main triggers for learning. In this sense, we realise that learning takes place in the environment in which we live. In this sense, Oliveira (1992, p. 24) states that for Vygotsky "culture becomes part of human nature in a historical process that, throughout the development of the species and the individual, moulds the psychological functioning of man".

Based on this assumption, the human cognitive structure arises from the social and cultural experience that the individual expresses through language and socially acquired and constructed behaviour.

In this period, the influence of two other great psychologists grew: the Belarusian Vygotsky and the Frenchman Henri Wallon (1879-1962), who spoke explicitly about the importance of interpersonal relationships and language.

In Wallon's view, humans are naturally, biologically, social, and the individual "is the meeting point between the organic and the social". In line with Piaget, these two scholars can be considered the founders of developmental psychology, with their focus on child development.

For Vygotsky, what makes us human is the ability to use symbolic instruments to complement activities, which have a biological basis. What makes us human, in Vygotsky's view, is our ability to imagine.

[...] when we learn the specific language of our sociocultural environment, we radically transform the course of our own development. Thus we can see that Vygotsky's vision gives importance to the social, interpersonal dimension in the construction of the psychological subject.

Of all the concepts developed by Vygotsky, the zone of proximal development (ZDP) is the one that has exerted the most influence on educational research and practice. The (ZDP) in short is what the child does today together with others, they will be able to do on their own tomorrow.

Still according to Vygotsky, schools are guilty of either proposing activities outside the limits of the ZDP (too many abstract concepts and requirements) or not taking its existence into account (as in the case of teaching based only on materials and waiting for the child to be "ready" to learn more sophisticated content).

2.5 . GEOGRAPHY TEACHING IN CONTEMPORARY TIMES: NEW DEMANDS AND INNOVATIVE DIDACTIC POSSIBILITIES

2.5.1. Education for the 21st century

Twenty-first century education is based on four fundamental pillars, which according to

Delors (2010, p. 31-32) states that the concept of lifelong education therefore appears to be one of the keys to the twenty-first century. It overcomes the traditional distinction between lifelong learning, enabling responses to the challenges triggered by a rapidly changing world. This is not new, since previous reports on education have emphasised the need to return to school to face new situations that arise in both private and professional life. The four pillars are thus constituted: Learning to know (acquiring tools for understanding), Learning to do (in order to be able to act on the environment), Learning to live together (cooperation with others in all human activities), and finally Learning to be (the main concept that integrates all the previous concepts).

Given the four fundamental pillars of education mentioned above, therein lies the challenge of being a teacher at this time, as it requires this professional to take ownership of these concepts and put them into practice in their daily classroom life. The goal of education must take into account all the potential of each individual, such as memory, reasoning, aesthetic sense, physical abilities and the ability to communicate.

This is how education is conceived, forming the citizen as a whole, which can be understood and defined when drawing up education programmes and new pedagogical policies. In this sense, forming complete citizens in accordance with the four pillars of education, capable of sensitising themselves to the various situations of everyday life, is something that the world lacks in human beings.

Still from this perspective, Morin (2003) says that our psychic world (dreams, ideas, desires, fears) is infiltrated by our vision of the outside world. He also points out that memory itself (selective) can be the source of many errors. In this way, closed theories in search of a single absolute truth would not be able to see their own errors. These would be the errors of reason, which confuses rationality (open, negotiates with irrationality) and rationalisation (refuses to challenge arguments).

In the Western world, individuals know their pensions and act through opposing pairs such as light/dark, good/bad, body/soul (unlike cultures that see these pairs as complementary, for example, the Eastern representation of yin/yang). Different cultures choose their beliefs and convictions using paradigms that determine cognitive stereotypes, which would be their cultural imprinting and normalisation (norms, prohibitions, blockages). These ideas are so strong that they end up possessing us, rather than being directed by us as human subjects.

The author suggests that, in the search for truth, we allow ourselves to be possessed by ideas of criticism, self-criticism, openness and complexity. The principles of pertinent knowledge: how to organise and articulate so much information? How can we understand the context and the complex (global) in the Planetary Age of this new millennium? Morin proposes a reform of thinking, in which we must make both particular contexts and the global and complex evident:

We need to recompose the whole in order to know the parts. Hyper-specialisation leads to a disjunction, a breaking of links, in which the global and the essential are lost. The gap between one specialisation and another hides the unexpected, the new and invention. Teaching the human condition. All of our conditions on planet Earth: cosmic, physical, terrestrial and human must be analysed in order to understand the interrelationship (human unity) between them and how the result of this conjunction has allowed man to differentiate himself from others through culture (human diversity).

In this context, the author addresses the complex human being, his antagonistic character, even recognising the value of madness: "Genius springs up in the breach of the uncontrollable, precisely where madness prowls". Teaching earthly identity in the 21st century (the planetary era requires polycentric thinking nourished by the cultures of the world). In contrast to the last century, which left us a legacy of death (wars, exterminations, nuclear weapons, ecological death, new and more resistant viruses and bacteria, increased drug use) and even the death of modernity (as unconditional faith in progress), we must now have 'earthly citizenship' as the new mission of education.

Morin's proposal for thought reform is based on the Earth as the first and last homeland and earthly civic consciousness: responsibility and solidarity with the children of the Earth, in other words, "symbiosophy" (the wisdom of living together). The 20th century discovered the loss of the future and its unpredictability: "the future is called uncertainty". History no longer repeats itself or progresses in a straight line. History is a complex of order, disinheritance and organisation. Old certainties are no longer useful and the uncertainty of knowledge (illusion and error) prevails.

In this case, in the field of action, one must know how to plan and predict strategies, counting on randomness, chance, the unforeseen and possible transformations, in a prudent and audacious way at the same time. Teaching understanding ("the ethics of understanding requires understanding incomprehension" p.99) Nowadays, despite almost total communication, it seems that human incomprehension is widespread. Comparing intellectual and human understanding, he states that the latter goes beyond explanation, as it is made up of subject-to-subject knowledge.

The obstacles to understanding would be, for example, egocentrism, ethnocentrism and reductive thinking. The solution? Tolerance, guaranteed by democracy, even if it means suffering to accept ideas (not insults and aggression) that oppose our own. For the author, to understand is to learn and relearn incessantly; and it is fundamental for the reform of mentalities. The ethics of humankind ("the imperative has become to save humanity by realising it") Individual/society/species are not only inseparable, but co-producers of each other. According to Morin, only democracy allows for consensus (respecting minorities), diversity and is nourished by the ideal of Freedom, Equality and Fraternity.

2.6   INNOVATIVE TEACHING

2.6.1. The geography teacher's practice

A dynamic and modern geography education needs to go beyond the classroom, beyond the school walls. According to Castrogiovanni (2009, p.13), "there is still little contact between the school and life, with the students' daily lives". The outside world always seems to be more attractive and dynamic than the internal, homogeneous space of the school. Outside, life is pulsating, happening, people are searching for their identity, with their conflicts, in their own spaces and times.

Does the geographical practice of contemporary teachers make it possible to read the world, think about and understand the space in which we live?

To answer this question, we need to understand what model of geography teacher is needed to guide geography education on the road to learning.

We will mention two models of teaching discussed by thinkers such as: Comenius, Roussean, Pestalozzi and Herbat. These models are known as: traditional pedagogy and renewal pedagogy.

The pedagogical ideas of these thinkers formed the basis of European pedagogical thinking (experimental method) and spread throughout the world, demarcating the pedagogical conceptions that today are known as traditional teaching and innovative teaching. For Libâneo:

> "Traditional pedagogy, in its various currents, characterises conceptions of education in which the action of external agents in the formation of the student predominates, the primacy of the object of knowledge, the transmission of knowledge constituted in tradition and in the great truths accumulated by humanity and a conception of teaching as an impression of images provided

With traditional classes, the teaching of geography becomes an uninteresting and discouraging subject, an element of a culture that needs memory to retain information about names, rivers, capitals, regions, countries, climate, relief, altitudes, etc. We know that the first decades of the 21st century saw the emergence of human geography, which goes beyond physical geography to study human beings and their relations of labour production with nature.

On the other hand, innovative teaching brings together currents that advocate school renewal, opposing traditional teaching. This pedagogy has the following characteristics: valuing the student, freedom to learn, freedom to express themselves, with their own initiatives stimulated by discipline, becoming the subject of their learning, of their own development; scientific treatment of the educational process, taking into account the successive stages of biological and psychological development.

Geography teaching needs to provide meaningful learning opportunities for students, and for this to happen, the teacher's methodology must first undergo reflection and change. The teacher's practice must relate the content taught in the classroom to the students' everyday lives.

The PCNs (2008, pg. 55) state that: "the concepts elaborated to form the basis of the structures of the knowledge to be built in High School are all elements that can be discussed and data elaborated that are essential to understanding geographical facts". In this sense, there needs to be a reformulation of the geography curriculum that takes into account students' local peculiarities.

Respecting the student's experiential background, their unsystematic or casual or even systematic learning, is also the aim of geography teaching, as it is an interactive process. It is also important to discuss differences, giving students the opportunity to build their knowledge by surveying problems observed in the school and community in general. Geography trains citizens to look critically at the world.

Geography should be taught in such a way as to equip students to deal with spatiality and its multiple approaches. They must understand social life, reflecting on the different subjects, actions and transformations. As well as reflecting on space in the face of so many changes in society.

> The choice of concepts rather than watertight definitions is essential for
> structuring Geographical Science, which seeks to free itself from the concept

One of the many things a teacher has to think about in the classroom is the use of the textbook. Following the textbook mechanically, without looking at the reality around you, for example, doesn't seem to be the best way forward. Castrogiovanni (2009, p.137) criticises this formality by stating "that we should build knowledge based on content and not on a sequence proposed by textbooks, as this may not be the most sensible way". An attentive and questioning eye is needed to assess the real need and importance of the content listed in these veritable manuals of serial reproduction.

So what is the way forward for effective geography teaching that is grounded in reality? According to Cavalcanti (2010, p. 139), in the teaching/learning process there is a relationship between the subject (student in activity) and objects of knowledge (elaborated knowledge) under the direction of the teacher, who leads the subject's activity in relation to the object, so that they can build their knowledge. This does not disregard the student's reality; on the contrary, this interaction, the means of teaching, enables the student to organise themselves to better understand the world, their own reality and fully develop their critical sense.

2.7   THE SCHOOL ENVIRONMENT AS AN OPPORTUNITY TO ACQUIRE KNOWLEDGE

The school must create positive conditions that encourage students to ask questions, create solutions, raise possibilities, develop hypotheses, explain their reasoning and experiment by testing their conclusions. From this perspective, the teacher's role will be to organise and personalise their educational projects and teaching methodology. In this way, they encourage autonomy, responsibility and student interaction so that they can acquire new knowledge, carry out new investigations and explore new stimuli based on the content they have learnt. In order to carry out an activity of this magnitude, students need to understand the other disciplines involved in the project, so interdisciplinarity comes into play.

As a theoretical foundation, Freinet's pedagogical proposal is divided into three sections:

> Experimentation, whenever this is possible, which can be both observation, comparison and control, as well as proof, through school material, of the problems that the mind formulates and the laws that it supposes or imagines. Creation, which, starting from the real, from instinctive or formal knowledge generated by conscious or unconscious experimentation, rises, with the help of imagination, to an ideal conception of the future it serves. Finally, completing, supporting and reinforcing them, *documentation* - the search for the desired information in different sources - which is like an awareness of the experience realised, in time and space, by other men, other races, other generations (Freinet, 1998, p. 354-355).

In the Freinet method, the school is advocated as a space linked to life, where the process of education must have a social meaning, where free expression is considered through pedagogy, where experimentation is the main axis on which all the students' acquisitions revolve. As teachers, our job is to create new needs in order to stimulate their creativity in acquiring new knowledge.

2.7.1. Overview of Freinet's Techniques

Pedagogical practices urgently need to be rethought because teaching is becoming increasingly uninteresting for students. Teachers need to update themselves and understand or look for new ways of learning that are meaningful in the student's life and that attract their attention.

Freinet's pedagogy is structured around a set of inseparable techniques that take the form of cooperative organisation. The classroom is a space for dialogue, choices and sharing knowledge.

The school newspaper, the wall newspaper, the conversation circle, inter-school correspondence, the book of life, the binder, the class album and the class outing are some of the Freinet techniques. The school newspaper must be realised by following the entire process that involves it. Over time, the choice of news to be published in the newspaper becomes increasingly democratic, prioritising the most interesting. Children look forward to drawing pictures, which are incorporated into news illustrations.

The conversation circle is the first time the class meets, it's a moment of free expression and each student has the opportunity to express their ideas, opinions and feelings. It's also a time to plan the day, discuss the content to be worked on and share news. The final round provides an opportunity to evaluate the activities carried out. It is a privileged moment for recording and systematising learning.

Inter-school correspondence takes place throughout the school year. This technique allows children to use different types of language to express and communicate their ideas, desires, curiosities and studies, so they use drawing, music, writing, poetry, painting, etc. By writing and reading the letters, the children are challenged and carry out constant research and investigations into natural phenomena, the school environment, neighbouring places, neighbourhoods, the family environment and the geographical environment. It is a means of publicising the albums and exchanging information with other children about the studies being carried out. Correspondence contributes to the appropriation of written and oral language by the children, who actively participate in every moment of the letter-writing process. The letters become reference materials for the children as they are displayed in the room.

The life book is a record of the most significant events in the class. In it, the teacher and/or students insert texts produced in the class or can record an important event that took place in the class or outside of it, such as an outing, a visit, a significant activity experienced by the student, the group of students, the family and the community. This record becomes a class diary over the course of the year, illustrated with drawings, photographs, reports and testimonies, which become part of the group's memory.

The class album is a collection of materials produced by the children on a particular subject that arouses their interest. Through individual or collective investigations, the children learn about their surroundings, their history, geography, people, community organisation, customs and the characteristics of the environment.

The cooperative school binder or document binder is an alternative form of teaching material, characterised by the organisation of specific subject sheets. This binder records the content studied by the class or a group of students.

Walking lessons are outdoors excursions that provide greater contact with the environment itself, allowing for discoveries that motivate the creation of free texts that can form part of the newspaper, the book of life or even inter-school correspondence.

Nowadays, due to the widespread use of mobile phones, the teacher can also use some applications that draw the students' attention to the lesson, and during the lesson, intersperse the use of cameras to record information and landscapes, so that the teacher is using what the students like with what they need to learn.

This Freinet technique or techniques are ideas that can be implemented in schools in a planned way and included in the school's PPP, as teachers used to work or develop activities in isolation, but the proposal is for the school to take ownership of the relevant knowledge and make it applicable in the classroom.

Teaching and learning in the digital world is also a proposal for teachers, with the aim of contextualising the experience of the Homos Zappiens student with the school curriculum offered to this student. From this perspective, teachers need to break old paradigms and appropriate new ones, using allied ICTs to improve the quality of the teaching and learning process.

Currently in this society known as the information society, where information is processed very quickly, as teachers we cannot omit or exclude the digital culture in which we are inserted, teaching and learning in the digital world requires practice, so that we have mastery or familiarity with the language and codes used in the digital world, and so that we can act in an updated way in the technological era, we have to seek knowledge and new learning, since such technologies change constantly, updating ourselves concomitantly with technological processes, we will have better conditions in dialogue with our student in everyday school life.

The proposal in question could be included in the school's Political Pedagogical Project (PPP), where the school will have the autonomy to reformulate or not through the webcurriculum. Learning and teaching require changes or the breaking of old paradigms and the acquisition of new ones, both for the teacher and the student, so the use of ICTs can be used as pedagogical tools that add to the quality of the teaching and learning process.

2.8   INTERDISCIPLINARITY AS A PROPOSAL FOR CHANGE

2.8.1 Origin and concept of interdisciplinarity

According to (Mendes; Vilela, 2003), it says that: from the 19th century onwards, with the emergence of industrialisation, and in order to meet the demand or need of that moment in industrial production and commercialisation, the process of specialisation was consolidated; specialisations in which each worker would have their own specialised function to carry out in the company.

Still on this subject, Carvalho (2002) starts with specialisation in industry, where everyone plays their part without knowing the whole process,
Thus, the model in force tends to be applied in the world of work, with specialised training. Known as the Taylorist and Fordist models, these technical, timed, standardised and routine work patterns are designed to increase productivity and reduce costs.

This model went into crisis in the 1970s, when the need to change or replace it became apparent. This post-Taylorism/Fordism replacement reconfigured labour relations into a flexible relationship known as Taylorism, where the worker accompanies production at all stages, initiating the process of globalisation.

The crisis of the 1980s and 1990s has come to an end, and now a new paradigm is beginning with the advent of globalisation, where the labour market is becoming more demanding due to technological advances, so the profile of the worker necessarily requires intellectual qualifications and preparation to keep up with the speed of change and adapt to current flexibility.

In this process of transformation or the replacement of paradigms, society also follows this movement, as do families and schools as elements that make up society. In this way, the pattern is followed by society and materialises in the fragmentation of knowledge through the emergence of disciplines, where each discipline will have a specialist, in other words, each teacher acts in their specific area in a fragmented way.

In secular life, human beings, in their incessant quest to satisfy their multiple and ever-historical needs of a biological, intellectual, cultural, affective and aesthetic nature, establish the most diverse social relationships, in other words, by nature, man has the need to have diverse relationships in order to satisfy his desires, needs, achieve answers and solutions. And this means that there is a multitude of relationships in his life. And what can we say about school life: this multi-relationship must exist in order to achieve the aims of systematic education?

In the course of the teacher's school practice, a certain isolation or lack of dialogue between knowledge is perceived, and with this knowledge becomes fragmented, from then on there is a need for a more holistic view of reality, but what does interdisciplinarity mean? What is its origin?

The author Paviani (2008) states in his work that the origin of interdisciplinarity lies in the transformation of ways of producing science and perceiving reality, and also in the development of the political-administrative aspects of teaching and research in scientific organisations and institutions.

The author meant that the concept of interdisciplinarity means that the geography teacher's pedagogical practice involves a relationship of reciprocity, of multidisciplinarity, which presupposes a different attitude towards the problem of knowledge. The action requires the geography teacher to make a permanent rational and critical effort. Paviani (2008) also says: "There are no formulas or models for interdisciplinarity"

What are the principles of interdisciplinarity that we should consider in the process of teaching geography? Based on studies by theorist Jayme Paviani, the principles are: "Unity and multiplicity;

continuity and discontinuity; complexity and emergence". (Paviavi, 2008). We quote:

Principle 1: Unity and multiplicity - The aim of interdisciplinarity is not to diminish or remove the specificity of sciences and disciplines, but to enable common relationships in the exchange between knowledge and reality.

2nd Principle: Continuity and discontinuity - To say that scientific knowledge is discontinuous is to say that it progresses from fact to fact, from aspect to aspect. However, while science describes, analyses and explains events and phenomena in isolation, theoretical discourse, laws and theories are characterised by a certain continuity or universality.

3rd Principle: Complexity and emergence - The principle of complexity goes hand in hand with that of emergence. One leads to the other. Complexity is a concept that seeks to express the multiple faces of reality.

Interdisciplinarity has two sides to it: one side, let's say the good side, the one that positively brings good and true results. Paviani (2008) says: [...] "The good part allows for new results that would not be achieved without this common effort and, in this way, modifies the nature and function of traditional disciplines". The other side is linked to the part that is not good, let's say "the bad". Interdisciplinarity is the bringing together of researchers from outside. Although they work together, each one is dedicated solely to their specialisation.

2.7.2. Education for the 21st century in geography teaching

In the current situation, according to Santos (1994), which is known as the technical-scientific-informational environment, this information occurs in a globalised way, and because it is inserted in this moment, the ideal would be for the geography teacher to appropriate new resources that the moment requires. The teaching and learning process in the 21st century is still challenging, as it requires a break with old paradigms and the acquisition of new ones.

Thus, the web can be used as a means and tool to help acquire and expand the teaching-learning process, as well as applications that the student knows about. With these means used, students feel more like learning, since it is or is related to what they like to do, some of them, such as: google maps, youtube, google docs.

Therefore, this current scenario is justified by the speed with which information and transformations in society occur through the use of network technologies (RTs), such technologies enhancing the student's geographical knowledge process.

2.8.3 The teaching and learning process in geography

In today's information age or technical-scientific environment, the geography teacher cannot remain static in the face of the transformations that have taken place rapidly.

So the teacher needs to re-signify teaching so that the student feels part of this process. Castrogiovane (2007, p.43) says that: [...] the teacher must take advantage of the students' experiences by problematising and contextualising them, in this way there will be a

insertion of life into the school, where the school will actually be integrated into the student's life. In this way, students will enjoy their studies, since they are part of the process, thus becoming actors and authors of their own life story.

In the face of all this dynamism, reflecting on geographical space is no easy task, given that the geography teacher's job is to bring together elements of analysis and social practices, which Castrogiovane (2007, p.43) says are:

> "If it's a struggle to exercise citizenship...and as an analysis and practice, raise issues such as land use planning and occupation, the right to health and education, access to housing and land, the preservation and conservation of biodiversity and environmental quality, and the need for cultural and material sustainability. What a responsibility!" CASTROGIOVANE (2007, p.43)

From the author's perspective, it can be said that the geography teacher needs to take social responsibility and look for methodological strategies that can serve and interact with the different social groups, integrating them into society, so that when they interact, they realise themselves as people and professionals; where this realisation comes about through the students' daily life practices, hence the importance of reviewing the "theoretical-methodological" practices of this science in the school environment where this research was carried out.

Castrogiovane (2007, p.44) says that: "Geography is perhaps the discipline that works most with interdisciplinary practices, traversing a range of possibilities in the field of education".

With this further tribute to the science of geography, it is believed that the self-esteem of geography teachers will be raised. By feeling good, they will be able to encourage their students to seek solutions to the problems to which they belong, as well as their collective participation in the actions in which they live, thereby minimising their feelings of powerlessness in the face of the problems of their daily lives.

The teaching process cannot be understood merely as the transmission of knowledge, the reproduction of knowledge.

Today we are talking about a new pedagogy that requires the teacher to have the skills to construct and reconstruct knowledge, as well as the student. And in the teaching-learning process, they are the subject of their own learning.

Morin (2000) defends this pedagogy of learning how to learn, where the teacher must organise learning environments to enable the student to become an autonomous subject in the process of building knowledge.

Students must be encouraged to develop their intelligence, build their knowledge and solve problems. Morin says:

> "The more developed general intelligence is, the greater its capacity to deal with special problems. Education must favour the mind's natural aptitude for posing and solving problems and, correlatively, stimulate the full use of general intelligence." (MORIN, 2003, p.18)

The author says that in order to favour the development of general intelligence, there needs to be exercise linked to doubt, i.e. the teacher must stimulate and arouse the student's curiosity so that they can acquire meaningful learning through problem-solving.

2.8.4 The various types of language in geography teaching

According to the PCNs 2006, the languages are: cartographic, textual, corporal and scenic, iconographic and oral:

> "Geography should encourage the reading of landscapes and maps as a teaching methodology so that students, in an innovative pedagogical practice, can observe, describe, compare and analyse the phenomena observed in reality, developing more complex intellectual skills" (PCNs, 2006, p. 51).

Languages serve as a support for geography lessons, i.e. they are a more suitable instrument for reading the geographical environment and its use, which involves the exercise of interdisciplinarity. Let's take a look at the types of language:

1° Cartographic language: Allows the student to make or draw up metal maps, as well as reading and using plan and thematic maps. And it also allows the use of ICT in geography, through aerial photography, digital maps and remote sensing, such technologies will enable a precise level of study, with regard to the localisation of spaces.

This acquisition of knowledge through cartography makes it possible to read and understand the place where you live. With mental maps, you know where you live, such as leisure areas, supermarkets, schools, etc, which are part of the student's day-to-day life.

2nd Textual language: Once they have grasped the basic concepts of geography, they will be able to produce texts based on observations of the landscape.

3° Body and stage language: Students will act out themes from their own reality through theatre plays.

3° Iconic language: Iconic language in geography also develops or sharpens students' skills through images, paintings, works of art, which is therefore a way of interpreting and describing a phenomenon by exposing what is observed in the phenomenon.

2.8.5. Text production in geography teaching using multiple languages

In all areas of knowledge there is a need to develop students' writing skills, because throughout their lives they will always be asked to relate to and produce texts. Thus, by appropriating the multiple languages already mentioned in this article, teachers can apply or use other languages from the student's everyday life in their classroom, such as: "A literary text, including poetry, an area photograph, or a film must meet the needs of geography teaching." (Filizola, 2009, p. 88)

According to this thinking, it can be seen that this practice adds a new meaning to knowledge through these languages, teaching becomes more enjoyable and less tedious, and it also allows students to express their ideas, dreams and choices on various topics in a spontaneous way, and it is up to the teacher to coordinate these activities in a creative and attractive way, so that they can better evaluate the educational process, both for the teacher and the student.

As an example of the use of text production in geography teaching, teachers can include the development of scientific projects in their praxis, learning from discoveries and problems observed and identified in students' daily lives. And from this observation, they can investigate and find answers. During the investigation, it is necessary to produce texts such as: the problem posed, the justification, the objects, the theoretical basis, the hypotheses, the methodological procedures, the discussion and results, and the conclusion. So, based on this practice, we're going to talk about the project methodology as a form of textual production for the teaching and learning process in geography.

According to the theorist Celso Antunes, the project methodology is an excellent tool to be considered as a mechanism to be used as a subsidy for textual production, as it allows students to investigate, evaluate and interpret everyday phenomena and this methodology enables students to systematise the knowledge they acquire from this perspective in an orderly and organised manner.

> "It transforms the student into a discoverer of meanings, it teaches them to research, to present the report of their research, it stimulates co-operation and sociability and in many cases it offers the student the opportunity to choose what role they want to play in the team they are part of." (Antunes, 2007, p. 117)

According to Antunes (2007), the purpose of the project methodology is divided into objectives, duration, subjects and developed competences expected of the teacher, where the objectives allow the student to discover scientific methodology, contextualised with their reality and also facilitating the construction of knowledge; the duration depends on the time it will take to complete the activities; the subjects the project covers all subjects, i.e. this methodology with a project is favourable to interdisciplinary actions, which can be developed throughout the year with such a proposal.

The competences developed by the students range from their relationship with the topic and the people involved to their role as participants in the methodology.

The competences expected of the teacher, on the other hand, require active action for positive guidance and development.

The use of this methodology not only changes the student's way of thinking, but also transforms the teacher's methodological approach; by changing their methodological approach, assessment results are better, as it allows the teacher to assess the process of student participation as a whole.

The project methodology also allows content to be worked on in a contextualised way that: "contemplates the observation, analysis, comparison, interpretation, synthesis and evaluation of places, landscapes, territories and regions." (Filizola, 2007, p. 25)

Among the various explanations, this is yet another reason to work on textual production in geography classes, as the student has the freedom to choose, guided by the teacher, the one that is of interest to them and their reality, in line with the pedagogical proposals for carrying out work that aims to go beyond decontextualised and isolated content.

2.9 INNOVATION IN GEOGRAPHY TEACHING AND LEARNING

When people talk about innovation, the word technology immediately comes to mind. We think of computers as an example of innovation. But in the very etymology of the word, innovation derives from the Latin innovare, which simply means to incorporate, to bring in, to insert the new, the novelty. In this sense, innovation is characterised by combinations of resources that generate new products, new processes, new markets, new forms of organisation and new materials.

But what is innovation anyway? Could this term be used in schools?

According to Souza (2014, p. 285), in his article entitled Innovative Strategies for Traditional Teaching Methods - General Aspects.

The author says that the concept of innovation is a "break with the dominant paradigm, and is the advance in different areas and forms and alternatives of work that break with the traditional structure".

From this perspective, innovation is all about strategies that go beyond the conventional and lead students to interact in the construction of their own knowledge, and where these strategies become attractive and enjoyable. In innovative methodology, the teacher mediates and guides the teaching and learning process, while the student is the proactive subject in the construction of their knowledge.

Innovative lessons are more than necessary in the education process and even more relevant in geography teaching, since geography is a dynamic science and, as such, it is essential to use alternative and innovative strategies associated with or contextualised to the content taught in geography classes.

Education has to surprise, captivate, conquer, enchant, enthuse and seduce at all times, so that knowledge is built through activities that excite curiosity, imagination and creativity.

Saviani (1995, p. 30) says of educational innovation, the term is understood as "putting educational experience at the service of new purposes", that is, in order to innovate it is necessary to start from a methodological reflection initially and then question the purposes of the educational experience in order to improve teaching.

What are innovative teaching methods? Understanding that they are new challenges that are being imposed in current educational scenarios.

From this perspective, the research deals with project methodologies in the teaching of geography, as proposed to reinforce the teaching practice of secondary school teachers at Mineko Hayashida State School.

The project methodology strengthens the teaching of geography, as it allows students to consider a particular problem and find a solution to it, as well as providing meaning to their learning.

2.9. 1Classroom experience   using the project methodology at Mineko Hayashida state school

Based on the importance of producing texts in geography lessons, the research reports on the experience of the researcher and a group of students during geography lessons at Mineko Hayashida state school. The project methodology emphasises geography education through research at school.

The classroom project methodology took place in three (3) stages:

1ª stage: geography lessons with scientific methodology:

The first orientations took place during geography lessons, from the third to the beginning of the fourth bimester of this year, which will take the form of basic notions about how the project methodology project can be implemented, and the criterion adopted for this investigation is based on a focus group, i.e. when the sample universe is large, a group is chosen as the focus for the study.

The steps of the project methodology applied in the classroom follow the steps below:

1- Introduction, in this text as a compulsory part of a project introducing the work to the reader.

2- Literature survey, books, magazines and the internet. Essential, as it will provide the basis for the work.

3- Problem, what? All scientific work starts with a concern, a problem, and from there the investigation begins.

4- Hypothesis, assumption. You need views, which are assumptions you have about a particular problem.

5- Objectives, goal, end. As an essential element, the objectives are important, on which the success of the research is to be achieved.

6- Justification is the conviction that the research is fundamental, what for? The why of such work is stated, highlighting its importance.

7- Methodology, a detailed explanation of the research, how? No less important is the methodology, which explains how the project will be carried out.

2ª stage: visit to the indicated location - Jari River

After explaining these seven (7) steps, the students went to the field for a local visit, on the left bank of the Jari River, to analyse the landscape; they took photographs and in the classroom they systematised what they had observed.

3rd stage: drawing up a questionnaire and applying it to the riverside population

Also in the classroom, the students drew up questionnaires that were later applied to the area under study in the Central and Malvinas neighbourhoods of Laranjal do Jari. The students were also given an idea of how to create a theoretical reference, as well as the opportunity to learn about the types of citations that would be used in their reference.

It was time to go to the field, where the places to be observed and questionnaires applied were the left bank of the Jari River in the Central and Malvinas neighbourhoods of the city of Laranjal do Jari, where housing is inadequate because there is no urban planning, so the case study was carried out.After collecting data through observation and questionnaire application, data was tabulated and analysed.

CHAPTER 3

METHODOLOGICAL FRAMEWORK

3.1 Type of research

The approach chosen in this research is qualitative and quantitative, where the actions were developed from the application of a questionnaire to teachers and students about their methodological practices, with their practices in geography classes, so qualitative and quantitative research was applied in the search to compare, analyse and interpret the data obtained from the current study.

Given these problems, the Project Methodology approach was born out of the methodological aspirations that need to be implemented in the school curriculum and, more specifically, in the teaching and learning of Geography.

According to the author Alvarenga (2014, p.11), the approaches taken in this research were qualitative and quantitative, which she calls mixed approaches, meaning a combination of qualitative and quantitative information.

3.2 Expected level of knowledge

The level of research being carried out is descriptive, which according to Alvarenga (2014, p. 40-41), descriptive research is aimed at determining what the variables are like or how they manifest themselves in a given situation, what their most characteristic features are and tracing possible relationships between variables. And in line with Gil (2008, p.28), he says that: descriptive research has the primary objective of describing the characteristics of a given population or phenomenon or establishing relationships between variables. There are countless studies that can be classified under this heading and one of the most significant characteristics is the use of standardised data collection techniques.

This investigation should identify the causes or difficulties of the students using the comparative investigation method where they were analysed in two groups, the group receiving lessons in the traditional way and the group receiving lessons with innovative methodologies, project methodology in the teaching of geography.

3.3 Study design

The methodology used to carry out this research was primarily bibliographical readings focused on the content under study, which were relevant to understanding and the critical and reflective development of the data analysed.

In the second strand, specific field research is carried out with geography teachers and with students in the first year of secondary school, where this field research enables a relationship between theory and practice in the content taught in the areas of geography.

3.4 Description of the Population and Sample

3.4.1 Population

In order to minimise this problematic situation under study, the project was developed at Mineko Hayashida State School, which runs secondary, integrated and innovative secondary education.

The population that makes up the universe of this research at Mineko Hayashida State School is made up of 15 classes of 40 students in the morning shift (600 students), 14 classes of 40 students in the afternoon shift (560 students), and 10 classes of 30 students in the evening shift (300 students), giving a total of 1,460 students and a total teaching staff of 60 teachers, including 5 geography teachers, 3 of whom are morning shift teachers, 2 afternoon shift teachers and 1 evening shift teacher.

3.4.2 Sample

The group of teachers who actually took part in the research accounted for 8.4 per cent of the total of 60 secondary school teachers at Mineko Hayashida State School, i.e. five (5) participants in the research, all of whom had a Bachelor's degree in Geography, a Specialist degree in Geography and a Master's degree in Geography.

The contingent of students who took part in the research represented 1.6 per cent of the total of 1,460 high school students at Mineko Hayashida State School, i.e. two hundred and forty (240) students, i.e. distributed in six (6) classes with 40 students, from the first grade of the institution, and this contingent was divided into two groups:

1st Group: 29 students who received innovative lessons over the course of a school year, i.e. four consecutive terms, with a total workload of 80 Geography lessons;

The 29 students were divided into sub-groups: nine sub-groups containing 3 students and one sub-group with 2 students.

The table below explains the development of projects as alternative work with students during Geography lessons. (Table 1)

Table 1: Proposal - Stages for developing an innovative methodology.

Exa mple	Group: Experimental			
	Innovative Techniques			
	1º Bimestre	2º Bimestre	3º Bimestre	4º Bimestre
	Concept of scientific methodology;	Structure of a project;	Drawing up a questionnaire	Data collection;
	Bibliographical research;	Research plan;	Visiting the banks of the Jari River;	Tabulation;
		Report;	Application of questionnaires;	Presentation of the project.

Source: Author, 2015.

2nd Group: 211 students who received traditional lectures over the course of a school year, i.e. four consecutive terms, with a total workload of 80 Geography lessons;

Both groups were taught by the same teacher, the researcher of this study.

According to the author Alvarenga (2014, p. 67 and 68), sampling will be probabilistic with systematic extraction. It is defined as probabilistic because of the need to know the probability of selecting a universe to analyse in the sample. The systematic extraction technique requires knowing the number of individuals that will make up the sample.

3.5 Site of   the research

3.5. 1 Geographical location of   the state of Amapá

The state of Amapá is located in the northern region of Brazil, capital Macapá, bordered to the (east) by the Atlantic Ocean, to the (north) by French Guiana, to the (north-west) by Suriname and to the (west and south) by the state of Pará.

Figure 1: Map of the State of Amapá
Source: Google

1.1.2 Location of the Municipality of Laranjal do Jari - Amapá

The municipality of Laranjal do Jari is located 260 kilometres from the capital Macapá in the south of the state of Amapá, bordering the municipalities of Vitória do Jari, Pedra Branca do Amapari and Oiapoque, as well as the state of Pará and the countries of Suriname and French Guiana.

1.1.3 Mineko Hayashida State School

The school is located in the municipality of Laranjal do Jarí, in the south of the state of Amapá, at Avenida Tancredo Neves number 2960, Bairro Agreste, and its legalisation act is Ordinance No. 050/93, opinion 139/0, Resolution 061/07-CEE/AP. It was founded in 1992 by a teacher who immigrated from Japan, giving rise to the name Mineko Hayashida.

Photo 2, **figure 2**: Mineko Hayashida State School

Source: Google

3.6 Research period

The research took place in 2015, in line with the school calendar of the school being researched, starting in March of the current year and divided into four bimesters, where the activities were applied weekly, two classes, twenty classes per bimester and eighty classes annually.

3.7 Tools and techniques

To collect the data, observation techniques and a questionnaire were applied to the research participants, geography teachers and first-year high school students at Mineko Hayashida School.

As for the geography teachers, unstructured observation was used, which, according to Alvarenga (2014, p. 83), is where the researcher records what they spontaneously observe, avoiding the rigidity of recording only what they have in their observation guide, having a clear research objective to guide their investigation.

And according to Lakatos (2003, p. 190), observation is a data collection technique for obtaining information and using the senses to obtain certain aspects of reality. It consists not only of seeing and hearing, but also of examining facts or phenomena that one wishes to study.

As for the students, field observation was used, which according to Alvarenga (2014, p. 86), this technique refers to the observation of reality in the natural place of the facts, in the field of action.

And according to Lakatos (2003, p. 191) It is a basic element of scientific enquiry, used in field research. Observation helps the researcher to identify evidence and obtain proof of objectives that individuals are not aware of, but which may guide their behaviour. It plays an important role in observational processes, in the context of discovery, where it forces the investigator into direct contact with reality.

This observation took place in a team of students, who observed the phenomenon from different angles, and then drew up a description taking into account the various points of view.

After the observation, a questionnaire was used, which according to Lakatos (2003, p. 200-203), is a data collection instrument consisting of an ordered series of questions. The application had a "face-to-face modality" Alvarenga (2014, p. 77)

The questionnaires used had open and closed questions. These are also called free or unconstrained questions. Informants can answer freely, using their own language, and express their opinions.

It allows for more in-depth and precise investigations; however, it has some drawbacks: it makes it difficult for the informant to answer the questionnaire himself, who has to write it down, the tabulation process, statistical treatment and interpretation. The analysis is difficult, complex, tiring and time-consuming.

Closed or dichotomous questions were also used. Also known as limited or fixed alternatives, these are questions in which the respondent chooses between two options: yes and no for closed questions and justified for subjective questions.

The study was based on the perception of the students themselves; their feelings, their ideas, their conceptions and their explanations of what happened, as well as involving the students' teachers. All the factors that may have contributed to the school failure were analysed with the teachers and the students. The result of the study should lead to an intervention in the educational process, establishing, proposing and applying techniques that favour the school success of the group of students.

The questionnaires - containing ten closed and open questions - were analysed and the student questionnaires - containing seven closed and open questions - were also assessed.

3.8 Data collection

Based on the application of the questionnaire and the data presented by the participant group, as an innovative methodology in the teaching of geography, the data was tabulated and consequently the tabulation of the data collected was carried out.

3.9 Pilot test

According to Alvarenga (2014, p. 82). The research uses a pilot test as an instrument to test the questionnaire to be applied during the field research. The purpose of this instrument is to check that the questionnaire does not pose any difficulties, that all the questions are clear and not difficult to understand, and that there are no questions left over or missing. If there are any doubts, the participants will tick the questions that cause doubts, that are complex, or if, in the students' opinion, there are any missing questions or alternatives. If irregularities are detected, they will be corrected. It took place on 26 September 2015.

3.10 Ethical considerations

According to Alvarenga (2014, p. 62), research must protect the sources of information of the people consulted. The research will be confidential, without exposing the names of students and teachers taking part in the research, or the names of classes. People's real names will be replaced by numbers or codes.

CHAPTER 4

ANALYTICAL FRAMEWORK

4.1 ANALYSES AND RESULTS

4.1.1   Part A - Research subject: teacher

The aim of this chapter was to draw up a profile of the teachers consulted, describing their geography teacher training, length of time in the classroom, continuing training and their conception of geography teaching in contemporary times. The results of teaching practice in the production of teaching quality are emphasised here.

Questions 1, 2 and 3 aim to outline the profile of the geography teacher at Mineko Hayashida State School.

Teacher 1 has a Bachelor's degree in Geography, a Master's degree in Geography Teaching and has been working in the field for approximately five years.

Teacher 2 has a BA and BSc in Geography, no further training and has been teaching geography for approximately 10 years.

Teacher 3 has a Bachelor's degree in Geography, a specialisation in Geography Teaching Methodology and has been working in the field for around five years.

Teacher 4 has a Bachelor's degree in Geography, a specialisation in Geography Teaching Methodology and has been working in the field for approximately five years.

Teacher 5 has a Bachelor's degree in Geography, a specialisation in Geography Teaching Methodology and has been working in the field for approximately five years.

It was observed that 20% of the teachers had a master's degree, 60% specialised courses and 20% only had an undergraduate degree (**Table 2, Figure 3 - Graph 1**).

Table 2: Continuing training.

Courses	Number of teachers	Percentage (%)
No training	1	20
Specialisation	3	60

| Master | 1 | 20 |
| Total | 5 | 100 |

Source: Author, 2015.

Figure 3: Relative frequency of continuing education for geography teachers at Mineko Hayashida State High School in 2015.

Source: Author, 2015

In question 4, it was observed that all the teachers uniquely stated that they use the textbook as an auxiliary tool in Geography lessons.

Question 5 showed that the majority of teachers (3) use the textbook constantly in their lessons. While two (2) teachers said they used the textbook sporadically because they developed handouts for their classes.

In question 6, it was noted that all the teachers consulted said that in their classes there is a need to relate Geography to other areas of knowledge, because they believe that teaching is not a compartmentalisation and that Geography is not an isolated science.

Question 7 probes the respondents' perception of the concept of innovation in geography lessons.

Teacher 1: says that innovative lessons enable students to reflect and learn using information from their daily lives, correlating it with the subject being taught. Using technology and new technologies in addition to textbooks.

Teacher 2: says that innovative classes go beyond the idea that knowledge, in order to be scientific, needs to break away from common sense. On the contrary, the student relies mainly on prior knowledge, in other words, on lived knowledge, so that they can relate the local to the global, starting from the particular to the general.

Teacher 3: says that he understands innovative classes to be classes taught in a dynamic way using new technologies such as computers, the internet, TV news, examples from everyday life and so on.

Teacher 4: He says that innovative classes are classes that use Datashow and classes that make it possible to use the internet.

Teacher 5: Says he doesn't use innovative lessons because the school structure doesn't allow it.

Question 8 was designed to find out which books those consulted had read on innovative methodologies in geography teaching.

Teacher 1: says that he has read books on innovative methodologies, but at the time of the researcher's enquiry, he was unable to answer the titles of the literature he had read.

Teacher 2: says that he has read the book entitled: nine innovative lessons at university.

Teacher 3: says that the book he read is called: Alternatives in Teaching Didactics.

Teacher 4: He's read it but can't remember the title of the book.

Teacher 5: You haven't read any books on this subject yet.

The aim of question 9 was to find out about the pedagogical praxis of geography teachers at Mineko Hayasida State School.

Teacher 1: In my pedagogical practices, I try to build a debate about the content to be studied, starting from the student's experience, what they know about the subject.

Teacher 2: I always relate the subjects worked on to the local and global relationship.

Teacher 3: My practice is expositive classes, usually aided by the use of the computer with documentaries, slides, films and so on.

Teacher 4: In general, the practice I use in my classes is dialogue-based lectures and research on the internet.

Teacher 5: I try to do more lessons specifically using the textbook, sometimes enabling my students to research on the internet.

Question 10 discusses the pedagogical model used by those surveyed during their geography lessons.

Teacher 1: In most of my classes, I use the critical-reflective model as a way of developing the student-teacher relationship, although sometimes I use the traditional model.

Teacher 2: Even with so much technology, I'm more focused on the traditional libertarian method, since the students' indiscipline requires this method.

Teacher 3: I follow the traditional model and it's up to the teacher to use mechanisms to enrich their lessons in their daily lives.

Teacher 4: I follow the traditional model, sometimes using some technology.

Teacher 5: I follow the traditional model, but I give students the chance to search for knowledge.

4.1.2 Part   B1 - Research participant: student - from the classroom to the field trip

Based on the study carried out on the steps of the scientific method in Geography classes and its application in the classroom, with the aim of facilitating learning in Geography teaching in secondary schools and providing students with learning through this practice.

Photo 23, **figure 4,** shows the application of the project methodology in Geography lessons: attached

Source: Author, Laranjal do Jari - AP, 2015.

[a]This didactic transposition led students from the 1st grade of secondary school to realise that they needed to prove the geographical concepts they had been taught in class in their everyday lives, on the left bank of the Jari River, located in the municipality of Laranjal do Jari - Amapá.

Knowing that geographical concepts were born out of scientific research, the students went into the field. To prove it.

Using the method of observation, the students realised these problems: River silting, pollution, inadequate housing, rubbish dumped in the open, human waste thrown directly into the Jari River in the open.

From the field trip, the need arose to draw up a project to minimise the anthropic actions seen in the area.

With this idea in mind, the students visited the Central and Malvinas neighbourhoods to conduct questionnaires with residents, as shown in the following images:

Photo 4, Figure 5: Questionnaire application with riverside dwellers

Source: Author, Laranjal do Jari - AP, 20

Photo 5, **Figure 6**: Observation of the area surveyed

Source: Author, Laranjal do Jari - AP, 2015.

Photo 6, **Figure 7**: Site surveyed, Malvinas neighbourhood

Source: Author, Laranjal do Jari - AP, 2015.

Photo 7, **Figure 8**: Site surveyed, Morro dos Macacos
Source: Author, Laranjal do Jari - AP, 2015.

4.1.3 Part B2 - Research subject: student - moment in the classroom

55

The questionnaire below contributes to meeting the objectives set out in the article, through which it will be possible to solve the research problem.

The data tabulated from this stage will reveal the following result:

In question 1, it was observed that the students surveyed gave their ideas for a possible change in the reformulation of thinking, and how a teacher could use other methodologies to improve the quality of teaching and learning.

In line with Freinet, teachers need to appropriate methodologies that can be interesting in the student's life (Table 3, Graph 2).

Table 3: Critical understanding of geography teachers' praxis

1ª Question:		
Answers	**Number of Students**	**Percentage (%)**
There are classroom lessons only	8	27,6
Teacher explains content well	13	44,8
More dynamic lessons	4	23,8

Source: Author, 2015

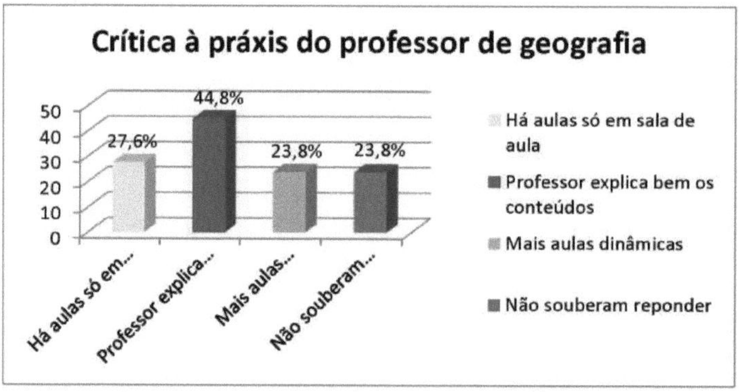

Graph 2: Critical understanding of the geography teacher's praxis
Source: Author, 2015.

This is how we have the data that confirms Freinet's technique: 27.6 per cent of those surveyed were emphatic in saying that lessons in the classroom alone become tiring, 44.8 per cent of the students said that the teacher explains geography lessons well, 23.8 per cent said that lessons need to be more dynamic and only 23.8 per cent of those surveyed were unable to answer.

In question 2:

One of the great challenges for geography teachers, according to the PCNs (2006, p. 46), is to create or formulate strategies that can attract students to the new teaching and learning in contemporary times (Table 4, Figure 10, Graph 3).

Table 4: Pedagogical models in the long history of society.

2ª Question:		
Answers	**Number of Students**	**Percentage (%)**
Couldn't answer	5	17
Today's lessons are more practical and contextualised	11	38
More up-to-date and dynamic teachers	4	14
In the past, with memorised lessons	9	31
Total	**29**	**100**

Source: Author, 2015.

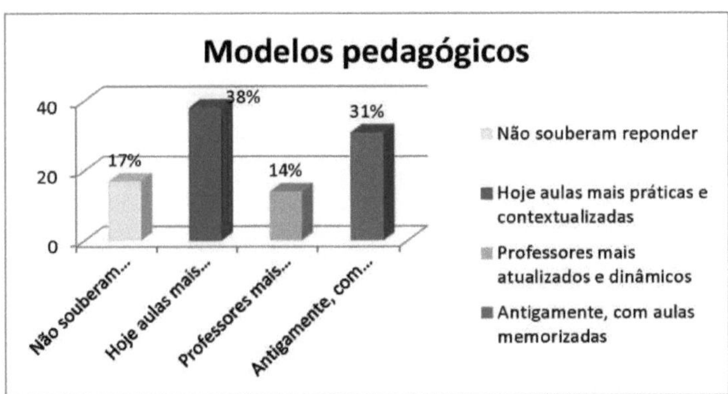

Graph 3: Pedagogical models throughout the history of society.
Source: Author, 2015.

17 per cent of students were unable to answer;

38% of the students said that classes in the past were traditional and nowadays they are more practical and contextualised;

14% of the students said that contemporary teachers are more up-to-date and dynamic;

31% of the students said that in the past lessons were more about memorisation and that today they are more contextualised;

Here we can see that traditional teaching is no longer interesting and in the 21st century we have to keep up with technological developments and update ourselves as education professionals.

In question 3:

These would be suggestions given by the students for the teacher to create methodologies that meet the learning needs of geography in a less complicated way, in line with what is recommended by the ENEM, since these students will be doing it (Table 5, Figure 11, Graph 4).

Table 5: Enabling inferences about new teaching practices

3ª Question:		
Answers	**Number of Students**	**Percentage (%)**
Teachers should do more practical lessons	18	62
No answer	11	38
Total	**29**	**100**

Source: Author, 2015.

Graph 4: Enabling inferences about new teaching practices
Source: Author, 2015.

62 per cent of the students said that the teacher should give practical lessons twice a week,

with video lessons and self-explanatory handouts and with activity books to prepare for the ENEM and with mock exams; 38 per cent of the students didn't know how to answer;

In question 4

Proving Freinet's technique through project pedagogy, the 29 students unanimously confirmed that lessons that go beyond the classroom, such as multimedia and practical field lessons, facilitate learning.

In question 5, using the Freinet technique, the 29 students stated that this way of teaching geography makes it easier to conceive learning.

In question 6, 59% of the students surveyed said that observation classes enable them to better relate theory to practice, and that it is easier when they visualise reality outside the classroom, while 41% of the students did not give a satisfactory answer.

We noticed that through the Freinet technique, knowledge becomes contextualised with practical theory and the reality experienced by the student (Table 6, Figure 12, Graph 5).

Table 6: Field lessons and observation improve the teaching and learning process

6ª Question: Field lessons and observation		
Answers	**Number of students**	**Percentage (%)**
Practical lessons facilitate and improve learning	17	59
No answer	12	41
Total	**29**	**100**

Source: Author, 2015.

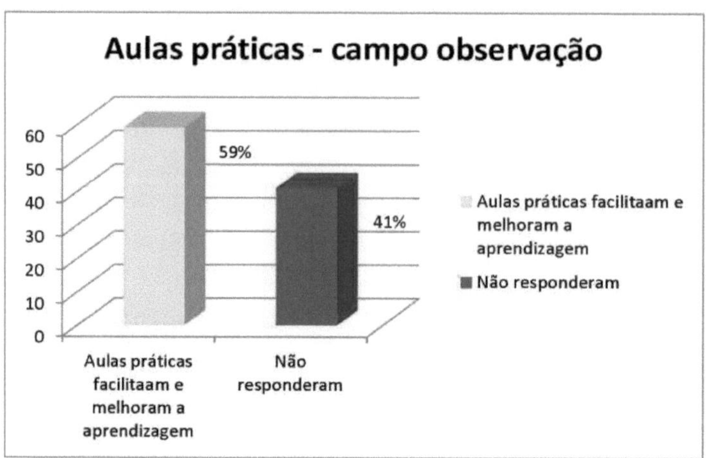

Graph 5: Field lessons and observation improve the teaching and learning process

Source: Author, 2015.

In question 7, the 29 students unanimously liked the field classes, because according to them, in this class they were able to relate theory and practice, so they were able to assimilate the subject seen in class more easily.

4.2 DISCUSSIONS OF THE DATA COLLECTED

4.2.1 Discussion of the data collected - Participant Subject: Teachers

This research shows that all the teachers consulted have a degree in Geography and a Bachelor's degree in Geography, which means that Mineko Hayashida State School has a staff of professionals able to teach high school geography classes.

Talking about academic training is important because, whether it is excellent or not, it reflects on the way we teach, which in turn affects the quality of teaching and consequently student learning. According to Callai (2013), public universities (especially the older ones) generally try to train bachelors in Geography, with a degree as a complement. This is done at the end of the bachelor's degree course, and it is up to the student to study only the didactic subjects. In this sense, their training may be inefficient compared to a full degree course in geography.

Cavalcanti (2002, p. 21) emphasises that:

> the teacher training process aims to develop critical-reflective competence, which provides them with the means for autonomous thinking, which facilitates the dynamics of self-education, and which enables the articulation of theory and teaching practice.

These are five teachers who have been trained in geography at school, not just in the science itself, but in how to administer lessons in a contextualised, dynamic way, linking subjects to students' everyday lives. Geography lessons need to be renewed in order to improve the teaching process and produce meaningful learning. Vieira & Sá (2007, p. 102) emphasise that:

> A dynamic classroom, in which the student participates as a subject in the shared construction of knowledge, can be very productive because the student is motivated to seek out information and is committed to the task.
> analyses to prove their arguments. It's a lesson rich in content and everyone leaves with improved knowledge, because co-operation in building collective knowledge motivates everyone who takes part. It's not reproduction, it's not "dictation", it's not copying: it's the authors' invention.

With regard to length of experience, as mentioned in question 2, it can be seen that the majority of the teachers (4) consulted have been working in the geography classroom for between 1 and 5 years, while the minority (1) has been working for between 5 and 10 years. If time is a variable proportional to the amount of professional experience, it can be said that only 20% of the teachers have been working for 10 years, mathematically speaking.

But in terms of teaching quality, do these 20 per cent perform better in the geographical teaching-learning process than the 80 per cent who have up to five years' experience in the classroom?

To answer the questions, teachers' length of experience alone does not mean that their lessons are better than those of other teachers. In order to maximise the quality of teaching, in addition to the time factor, it requires continuing training, teaching resources (books, magazines, multimedia resources) and a physical structure.

Analysing question 3 in particular, which deals with the continuing education of geography teachers at Mineko Hayashida State School, we comment:

A state of formation suggests growth, development. Education professionals must be constantly on the move. In line with Lima (2002, p. 244) when he states that a:

Continuing education has as its starting and finishing points the process of articulation between competent and reflective teaching work, being conceived as the process of articulation between teaching work, knowledge and the teacher's professional development, as a possibility of a reflective posture energised by praxis.

According to the statistical data collected in the survey, it was found that the teachers involved in the research had 20% master's degrees, 60% of those consulted had completed specialisation courses and 20% had ended their academic life and had not continued their training.

And emphasising the 20% corresponding to the lack of continuity in geography teacher training, he argues that educating requires understanding that man is a totally changeable being, and that he is the result of experiences, of the relationships he establishes with others. Man is capable of change, and this change favours the way he educates, which in turn favours the way he educates.

This is why, when they leave university, graduates should not think that their academic life is over. In this sense, professionals must be constantly updating the skills inherent in their profession.

Questions 4 and 5, which belong to the questionnaire applied to the teachers, deal with the use of textbooks in geography classes. It can be seen that the textbook culture is strongly present in the school environment, especially in geography classes, as the main teaching tool, which often, due to the physical structural condition of Mineko Hayashida State School, becomes the only tool to help high school students learn to understand the space where they live geographically and the relationship between that space and other people.

Lopes (2007, p. 208) gives a classic definition of a textbook as "being a didactic version of knowledge for school purposes and/or for the purpose of forming values". In dialogue with Lopes, the culture of the textbook is ancient, but it should not be a finished end in itself, but one of the means of contributing to student knowledge. Emphasising that nowadays there are other interactive ways of acquiring knowledge, for example through computerised technologies, among many others.

With regard to question 6, there was a unity in the teachers' answers, stating that there is a need to relate geography to other sciences in their classes, as they believe that geography is not an isolated science. This interdisciplinarity allows or opens up a range of possibilities for intervention or relating to the quality of the learning process. According to Auler (2007, p.7), "interdisciplinarity requires analysis from various disciplinary perspectives articulated around a theme consisting of an open problem, with environmental problems being typical representatives." In this respect, it was possible to see the relationship that the teachers made when running classes with other sciences and content.

With regard to question 7, there are some answers regarding teachers 1, 2, 3 and 4's conception of innovative teaching classes. They say that innovative lessons enable students to: reflect and learn; convert unsystematic knowledge into systematic knowledge; use new technologies in a dynamic way. Teacher 5 says that his classes are not innovative because of the structure of Mineko Hayashida State School, which does not allow it.

From these answers, it can be seen that the teachers' concept of innovative lessons is linked to the use of technology, which includes computers, mobile phones, projectors, the internet, TV and many other things. Among the answers, one that went even further was emphasised: "the physical structure of the school that allows or doesn't allow innovative lessons to be developed and carried out", in other words, if there isn't a good school structure, the lessons will be merely theoretical, without "something new" to attract students' attention and encourage them to take part in geography lessons with questions about the content and statements about what they contemplate in their daily lives, which become a curiosity or even a doubt.

The concept of school innovation in geography lessons goes beyond the use of technology. This innovation can be "something new", for example: poetry, theatre, music, videos, visiting parts of your city that arouse interest in geography lessons and creativity in the acquisition of knowledge. Cardoso (2003), based on his studies, presents the concepts of novelty, change, process and improvement as essential attributes for defining innovation.

According to Souza (2014, p. 285), in his article entitled Innovative Strategies for Traditional Teaching Methods - General Aspects.

The author says that the concept of innovation is a "break with the dominant paradigm, and is the advance in different areas and forms and alternatives of work that break with the traditional structure".

From this perspective, innovation is about strategies that go beyond the conventional and lead students to interact in the construction of their own knowledge, and where these strategies become attractive and enjoyable. In innovative methodology, the teacher mediates and guides the teaching and learning process, while the student is the proactive subject in the construction of their knowledge.

Innovative lessons are more than necessary in the education process and even more relevant in geography teaching, since geography is a dynamic science and, as such, it is essential to use alternative and innovative strategies associated with or contextualised to the content taught in geography classes.

For question 8, the survey showed that two teachers said they had already read books on didactics or innovative methodologies, but at the time the questionnaire was administered, they were unable to name the titles of the literature they had read. On the other hand, two teachers said they had read books entitled: Nine Innovative Lessons at University; Alternatives in Teaching Didactics. Finally, another teacher said that he had not yet read any books on this subject.

The results show that there is a lack of reading on didactic strategies; strategies which

Thus, it is extremely important for teachers to keep up to date with reading of this magnitude in order to enrich their knowledge and, at the same time, reflect on the quality of the student's learning. Kleiman (2007, p. 20) states that "the more textual knowledge the reader has, the greater their exposure to all types of text, the easier it will be to understand".

For question 9, it is of the utmost importance that teaching practice undergoes reflection and changes in attitudes as there is a need to improve teaching and provide meaningful learning.

According to Pimenta:

> Education is a social practice. But practice doesn't speak for itself. It requires a theoretical relationship with it. Pedagogy as a science (theory), when investigating education as a social practice, provides the "theoretical ingredients" necessary for knowledge and intervention in education (social practice) (2002, p. 93/94).

The content taught in the classroom needs to be interlinked with theory and practice in a way that is unquestionably consistent with the day-to-day school life experienced in the classroom. This means that theory and practice cannot be separated.

The teachers who took part in the survey recognise the importance of linking theory with practice. And in their speeches, they emphasise that geography teaching needs to be taught in a dynamic, contextualised and up-to-date way with connections to local, regional and global reality, using technological tools to facilitate learning.

For question 10, the responses of the teachers consulted were analysed as follows:

Teacher 1's response is mistaken, given that the critical-reflective process is an action within the ongoing educational process. He also uses traditional practices to teach his students.

Teacher 2 uses the traditional model, which according to Libâneo (1994, p.64) is a pedagogy in which the activity of teaching is centred on the teacher who explains and interprets the subject. And in this traditional practice, the teacher also uses libertarian pedagogy, which according to Libâneo (1994, p.69), states that:

> It is centred on the discussion of social and political issues; one could speak of teaching centred on social reality, in which the teacher and students analyse problems and realities of the socio-economic and cultural environment, of the local community.

Based on the response of teacher 2, it is believed that in his teaching practice, he uses the traditional model, but initiates and leads classroom discussions during geography lessons about his students' social issues.

Teachers 3, 4 and 5 tend towards traditional pedagogy. According to Libanêo (1994, p. 64), "in traditional pedagogy, didactics is a normative discipline, a set of principles and rules that regulate teaching. The teaching activity is centred on the teacher who explains and interprets the subject".

Traditional didactics have stood the test of time and continue to prevail in schools. It is common for schools to assign teaching the task of merely transmitting knowledge without

gives them the chance to reflect on what is being taught, what is its importance and what is its relationship with other sciences.

The conclusion is that the teaching practice of those surveyed is still based on traditional didactics and that these teachers, despite the traditional method, are concerned with teaching not through repetition and memorisation, but through action.

4.2.2 Discussion of the data collected - Participants: Students

1ª) Time to visit the seafront - Field lesson

This research found that the field lesson, a visit to the lower part of the city, located on the left bank of the Jari River, in the central neighbourhood of the city of Laranjal do Jari - Amapá, became an instrument that facilitated learning in geography for the students of the 1stª grade of secondary school at Mineko Hayashida State School.

Throughout this research, the importance of innovating geography classes has been supported, so that it enables students to think critically and holistically about their reality. However, here in this research, the difficulties faced by the teachers participating in the research in transforming their classes into environments conducive to learning with classes that are meaningful to students' everyday lives have materialised.

Before the visit, geographical concepts were taught in the classroom through a scientific investigation. These geographical concepts required the students to come up with some answers, for example: what anthropogenic actions have led to the siltation and pollution of the Jari River? In order to find the answers, they were taught in class how to investigate using the steps of scientific methodology, which are: problem statement, objectives, justification, hypotheses, methodology and conclusion.

The class of 29 students was divided into groups to facilitate interaction between students, so that together they could find solutions. From this training and notions of initiation, of scientific lessons, they went on to prove their hypotheses raised in class.

In short, the aim of the field class visit was to analyse the benefits of using the field class as a methodological tool for teaching geography. Carbonell (2002) points out that "spaces outside the classroom awaken the mind and the capacity to learn, because they are characterised as stimulating spaces which, if well used, are classified as a relevant setting for learning". And Oliveira (2009) emphasises that:

> Fieldwork makes it possible to get closer to reality, as contact with the phenomena presented in space leads to reflection in search of the essence, as it allows us to see physical and metaphysical characteristics that are obscured visually and intellectually in a frozen representation of the landscape, whether materialised by maps, photos or aerial and orbital images. Thus, not

2nd) Questionnaire application

According to the results of the questionnaire, the acquisition of significant learning as a benefit of the programme was proven.

use of field lessons. Only 29 students said that it was a pleasure to visit the banks of the Jari River in order to visualise in practice the geographical concepts taught in class.

It was also observed that the field lesson: aroused greater interest in geography lessons, greater understanding of the content studied in the classroom, with greater student/teacher and student/student participation and interaction, as well as awakening a critical sense and possible awareness of the perceived environmental and social issues in the place visited.

This research showed that the project methodology advocated by Freinet favours a better understanding of the content; This understanding is reflected in the quality of Geography teaching and learning, and according to the students, classes outside the school environment should take place more often and in other disciplines. From this perspective, interdisciplinary teaching is appropriate, given the current crisis in education, the way out of the problem would be for the school to work through the project methodology, since it favours interdisciplinary work and is in line with the PCNs (2008), where it "refers to problem solving from this interdisciplinary perspective".

Based on the above results, this article proposes a change in the curriculum to include project methodology in its PPP, thus enabling interdisciplinary and contextualised teaching and learning.

CHAPTER 5

CONCLUSIONS AND RECOMMENDATIONS

5.1 CONCLUSIONS

In general, the problem of traditional classes in the science of geography goes beyond the issues and infrastructure of public schools, the lack of continuing training for teachers is a constant in the educational environment, based on this assumption there is a low quality teaching and learning process.

Geography in itself is an interdisciplinary science. From this point of view, the process of teaching and learning geography needs a new look and a new professional attitude. Knowing that the current situation demands a new perspective from geography teachers, this is why the subject under study is necessary.

Answering the general objective of this research: "to analyse the innovative didactics used by teachers in the teaching of geography at secondary level, at Mineko Hayashida State School" in 2015, in the municipality of Laranjal do Jari, in the state of Amapá, Brazil.

It can be concluded that the only way to solve the problem under study is to introduce innovative, interdisciplinary lessons that appeal to students' interests, where they can act in a pleasurable way to build their own knowledge in a way that is meaningful to their lives.

And according to the first specific objective: to describe the praxis of high school geography teachers at Mineko Hayashida State School.

The conclusion is that the teaching practice of the teachers surveyed needs methodological reflection, such as participation in seminars, specific websites for teachers, participation in continuing education courses, as well as literature on geography teaching practice.

It concludes that: And according to the second specific objective: to demonstrate the methodological strategies by the geography teacher at Mineko Hayashida State School.

The author concludes that during the data collection through questionnaires and informal conversations with the teachers participating in the research, it was realised that there is no constant search on the part of the teachers for alternatives to overcome the difficulties faced by basic education teachers in the construction of geographical knowledge. And because the Mineko Hayashida State School doesn't have an excellent physical structure, the geography teacher is left to use the textbook in almost every lesson as the main teaching tool to aid the teaching and learning process.

And according to the third specific objective: to point out the need to apply new methodologies in the teaching of Geography at Mineko Hayashida State School.

The project methodology allows students to learn content without memorising it, thus enabling them to reflect on and analyse their knowledge. From this perspective, teaching geography becomes more attractive, enjoyable and meaningful for students.

He concludes that by analysing the data obtained, it was observed that the group made up of 29 students, accompanied by differentiated and methodologically innovative classes, lead students to participate more effectively in the teaching and learning process, in this perspective this group of students obtained a positive performance in the approval of 100% of the group participating in the study, while the opposite group, assisted by traditional classes made up of 211 participating students, did not have the same qualitative performance, since the traditional classes were not attractive and significant for this group.

The conclusion is that there is a need for reflexive interventions to break old paradigms and acquire new ones, providing effective development of the teaching and learning process, which consequently improves and increases the quality of geography teaching and learning in secondary schools. The study therefore sought to awaken students' interest in geography lessons, which many see as a merely "decorative" subject. With the field lesson, a tool that facilitates learning, the students were able to identify the geographical content taught in class and relate it to practice, and they were able to see, feel and, consequently, understand.

5.2 RECOMMENDATIONS

In view of the purpose of this research with teachers and students at Mineko Hayashida State School, it points to methodological reflection, changes in attitudes and a constant search for alternatives to improve geography teaching. In view of this need, we recommend

The Secretary of State for Education (SEED) of the State of Amapá Brazil, which administers education in the aforementioned state, should implement the reformulation of the geography teaching curriculum in conjunction with the state public schools that are subsidiaries of this higher body, inserting in their Political Pedagogical Projects (PPP) a revitalisation or implementation of the same, where the Mineko Hayashida State School could be used as the pilot school for this proposal and then expanded to other public schools throughout the state of Mapá;

F Continuous training for teachers, specifically geography teachers, school management and pedagogues, specifically in innovative methodologies that will improve teachers' praxis, considering pedagogical workshops in the school itself;

R Revitalisation or active use of the pedagogical political project (PPP) at Mineko Hayashida State School, in which it takes into account the collective participation of teachers, students, teachers, parents and school staff, i.e. the school community involved in this construction process.

A The inclusion of innovative activities (field classes, basic notions of scientific methodology) in geography teaching, which are included in the revitalisation of the PPP.

REFERENCES

Alvarenga, Estelbina Miranda de. (2014) *Metodologia da Investigação. 2ª* Auler, D. *Enfoque ciência-tecnologia-sociedade: pressupostos para o contexto brasileiro. Ciência & Ensino.* v. 1, p. 1-20, 2007.

Callai, Helena Copetti. *The formation of the Geography professional: (The teacher).* Ijuí: Unijuí, 2013.

Carbonell, J. *A aventura de inovar: a mudança na escola.* Porto Alegre: Artmed, 2002 (Pedagogical Innovation Collection).

Cardoso, A. P. O. *A Receptividade à Mudança e à Inovação Pedagógica:* o *professor e o contexto 10 escolar.* Porto. Edições Asa. 2003

Cavalcanti, Lana de Souza. *Geography and teaching practices.* Goiânia: Editora: Alternativa, 2002.

Célestin, Freinet/Lois Legrand; translation and organisation: José Gabriel Perissé. 150 p.: ill. - (Educators Collection). Brasília: Massangana, 2010. p. 15-17.

Cunha, A.G. *Dicionário etimológico Nova Fronteira da Língua Portuguesa.* Ed. Rio de Janeiro: Nova Fronteira, 1996.

edition. Graphic edition: A4 Disenos. Asunción - Paraguay.

Ferreira, A.B. de H. *Novo Dicionário da Língua Portuguesa.*Rio de Janeiro: Nova Fronteira, 1996.

Freinet, E. Nascimento de. *Uma Pedagogia Popular* - Lisbon: Editorial Estampa, 1969.

GIL, Antônio Carlos. *How to prepare research projects.* 6th ed. São Paulo: Atlas, 2008.

http://livrepensamento.com/2013/10/21/o-metodo-dialetico materialist/accessed: 10/11/2015.

http://portal.mec.gov.br/index.php PCNs

http://www.google.com. br. Accessed on 22/04/17

http://www.webartigos.com/artigos/os-pilares-de-jacques-delors/33899/

http://www3.fe.usp.br/secoes/inst/novo/agenda_eventos/inscricoes/PDF_S WF/14597.pdf accessed on 03/11/2015

Kleiman, Ângela. Text and reader: cognitive aspects of reading. 10. ed. Campinas: Pontes, 2007.

Lakatos, Eva Maria. Fundamentals of scientific methodology. 1 Marina de Andrade Marconi, Eva Maria Lakatos. - 5. ed. - São Paulo : Atlas, 2003.

Libâneo, José Carlos. Didactics. São Paulo: Cortez, 1994.

Lima, M. S. L. Supervised internship practices in continuing education. In: ROSA, D. E. G.; Souza, V. C. (Org.). Didáticas e práticas de ensino: interfaces com diferentes saberes e lugares formativos. Rio de Janeiro: DP&A, 2002. p. 244.

Lopes, Alice Casimiro. Curriculum and Epistemology. Ijuí: Editora Unijuí, 2007, p. 205-228.

Morin, Edgar. The well-made head: rethinking reform, reforming thought / Edgar Morin; translated by Eloá Jacobina. - 8th ed. -Rio de Janeiro: Bertrand Brasil, 2003.

Morin, Edgar. The seven knowledges necessary for the education of the future. 8 ed. São Paulo: Cortez; Brasília, DF: UNESCO, 2003.

Oliveira, A. M. M.; Miranda, S. L. ; Silva, T. D. ; Reis, L. . A Fieldwork Methodology for Teaching Geography in the Early Grades of Elementary School.In: XII Encuentro de geógrafos de América Latina, 2009, Montevideo.Caminando en una América Latina em trnasformación, 2009.

Paviani, Jayme. Interdisciplinarity: concepts and distinctions. Rev. ed. 2. Caxias do Sul: Educs, 2008.

Pimenta, Selma Garrido (Org). Didactics and teacher training: paths and perspectives in Brazil and Portugal. São Paulo: Cortez, 1996.

Rego, Nelson, Castrogiovanni, Antônio Carlos. Geography: pedagogical practices for secondary education - Porto Alegre: Artmed, 2007.

Saviani, D. A Filosofia da educação e o problema da inovação em educação. In: GARCIA, W. E. Inovação Educacional no Brasil: problemas e perspectivas. São Paulo, Cortez Editora, 1995.

Souza, et. Al. Innovative strategies for traditional teaching methods - *general aspects.* Revista da

Faculdade de Medicina de Ribeirão Pretoe do Hospital das Clínicas da FMRP Universidade de São Paulo, Campus Universitário - Ribeirão Preto - SP, 47(3):284-292, Jul.-Sept. 2014.

Vieira, C. E. & Sá, M. G. Teaching *resources: from the blackboard to the projector, what's changing?* In: Passini, E. Y. Prática de Ensino de Geografia e Estágio Supervisionado. São Paulo: Contexto, 200.p. 102.

Zotti, S.A. Curriculum. In: State University of Campinas. Faculty of Education. *Navigating the history of Brazilian education*, 2008.

Annex: Letter of acceptance

UNIVERSIDAD SAN LORENZO

CARTA DE APRESENTAÇÃO DO PROJETO DE PESQUISA

Paraguai, 22 de Novembro de 2015.

Apresentamos o(a) aluno(a) **ELEN SILVA DE ANDRADE**, portador(a) do RG nº **251821** e CPF nº **570.083.452-91**, regularmente matriculado(a) nesta Universidade, no curso de Mestrado em **EDUCAÇÃO**, para junto a essa instituição proceder sua pesquisa intitulada **DIDÁTICAS INOVADORA NO ENSINO DA GEOGRAFIA NO NÍVEL MÉDIO NA ESCOLA ESTADUAL MINEKO HAYASHIDA, LARANJAL DO JARI-AMAPÁ ,BR 2015** que será realizada com **Participante e Amostras**. A referida pesquisa tem como objetivo a realização da dissertação como exigência e pré-requisito para a conclusão do curso.

A Universidade San Lorenzo - UNISAL - Paraguai, vem por meio desta informar que a principal responsável pelo projeto de pesquisa é o(a) aluno(a) citado(a) que deverá se comprometer a utilizar todos os dados coletados, unicamente para o trabalho intitulado acima, bem como manter sob sigilo a identificação dos participantes, cujas informações teve acesso e, respeitando assim o código de ética desta universidade.

O(a) aluno(a) supracitado(a) deve comprometer-se a enviar ao seu orientador os documentos por ele solicitados, para acompanhamento da pesquisa. Também deverá enviar a carta de aceitação da instituição escolhida pela mesma para a execução de sua pesquisa, a partir da data determinada, que deverá ser informado da evolução da pesquisa ou de qualquer alteração que venha ocorrer imediatamente através de email.

Informamos, ainda, que todos os alunos desta instituição são devidamente orientados e acompanhados a fim de não infringir o código de ética, ficando vetada a utilização do nome da instituição para projetos pessoais. Sem mais para o momento.

Enelda Furtado França
Diretora
E. E. Mineko Hayashida
Decreto nº 0708/2015-GEA

Tutor Acadêmico

Appendix 1: Questionnaire applied to Geography teachers.

TITLE: INNOVATIVE DIDACTICS IN THE TEACHING OF GEOGRAPHY AT SECONDARY LEVEL MINEKO HAYASHIDA STATE SCHOOL, LARANJAL DO JARI-AP BR, 2015.

Dear interviewee:

The questionnaire below refers to field research for the defence of the dissertation of the researcher: Elen Silva de Andrade, from the Master's course in Educational Sciences, whose objective is to analyse the innovative didactics of the teacher in the subject of Geography in High School at Mineko Hayashida State School in Laranjal do Jari - Amapá.

Interviewee No. 01

Area: Grade 1 students [a]

Instructions:

1. **Only one alternative** should be ticked on all the questions below;
2. Use a **blue** or **black biros**;
3. Cannot be erased and/or left blank; (**no erasures; answered without exception**)
4. You do not need to identify yourself (**do not sign your name**).

Note:

Dear Teacher, it is of the utmost importance that you take part in this field research in order to obtain statistical data. I would like to emphasise that your collaboration in this questionnaire is spontaneous and free of charge. **QUESTIONS:**

1. What is your academic background for teaching geography?
a) (   ) Full degree in Geography.
b) (   ) Full Licence and Bachelor's Degree in Geography.

c) (   ) Bachelor of Geography.

d) (   ) Full degree in History.

2. How many years have you been a high school geography teacher?

a) (   ) 1 to 5 years.

b) (   ) 5 to 10 years.

c) (   ) 15 to 20 years.

d) (   ) 20 to 25 years.

3. Which area of geography do you have further training in?

a) Specialist in Geography Teaching Methodology

b) Masters in Geography

c) Masters in Education

d) Specialising in different areas

4. Do you use textbooks in your lessons?

a) (   ) Yes

b) (   ) No,

5. In your school where you work, do you use the textbook constantly?

a) (   ) Yes

b) (   ) No,

6. During your lessons, do you relate them to other sciences?

a) ( ) It is necessary to make this connection since knowledge is not compartmentalised, since classes should be interdisciplinary.

b) ( ) No, teachers should only focus on teaching geography.

7. What do you mean by innovative teaching lessons?

____

____

____

____

8. Have you read any literature on didactics or innovative methodologies?

a) (   ) Yes, which

(is)?____

b) (   )

No, ____

9. What is your practice as a high school geography teacher at Mineko Hayashida?

____

____

____

10. Which pedagogical model do you use in Geography at Mineko Hayashida?

____

____

____

____

ACKNOWLEDGEMENT:

Thank you for your participation in this stage of the research, as it was of the utmost importance for the realisation of this work.

Appendix 2: Questionnaire given to students in the 1st grade[a] .

TITLE: INNOVATIVE DIDACTICS IN THE TEACHING OF GEOGRAPHY AT SECONDARY LEVEL MINEKO HAYASHIDA STATE SCHOOL, LARANJAL DO JARI-AP BR, 2015.

Dear interviewee:

The questionnaire below refers to field research for the defence of the dissertation of the researcher: Elen Silva de Andrade, from the Master's course in Educational Sciences, whose objective is to analyse the innovative didactics of the teacher in the subject of Geography in High School at Mineko Hayashida State School in Laranjal do Jari - Amapá.

Interviewee No. 01

Area: Geography Teacher

Instructions:

1. **Only one alternative** should be ticked on all the questions below;
2. Use a **blue** or **black biros**;

3. Cannot be erased and/or left blank; (**no erasures; answered without exception**)

4. You do not need to identify yourself (**do not sign your name**).

Note:

        Dear Student, it is of the utmost importance that you take part in this field research in order to obtain statistical data. I would like to emphasise that your collaboration in this questionnaire is spontaneous and free of charge.

QUESTIONS:

1- What is the understanding of the critical form in the praxis of high school geography teachers?

2- Do you notice the differences in the way teachers teach throughout your school life?

3- How can we work on new geography teaching practices with a view to ENEM?

4- How would you like to learn geography?

a) Through traditional classroom lessons only.

b) Through multimedia-only lessons.

c) Through lessons that go beyond the classroom, such as multimedia and practical field lessons.

d) There's no way you like studying geography.

5- Should practical field/trip and observation lessons be used in other subjects? Justify your answer.

6- How can a field lesson/observation improve the teaching-learning process?

7- Did you enjoy the field lessons? Why? Were you able to learn more easily in this type of class?

ACKNOWLEDGEMENT:

Thank you for your participation in this stage of the research, as it was of the utmost importance for the realisation of this work.

Buy your books fast and straightforward online - at one of world's fastest growing online book stores! Environmentally sound due to Print-on-Demand technologies.

Buy your books online at
www.morebooks.shop

Kaufen Sie Ihre Bücher schnell und unkompliziert online – auf einer der am schnellsten wachsenden Buchhandelsplattformen weltweit! Dank Print-On-Demand umwelt- und ressourcenschonend produziert.

Bücher schneller online kaufen
www.morebooks.shop

Printed by Books on Demand GmbH, Norderstedt / Germany